U0315197

建筑邦 www.architbang.com

DESIGN

超 100,000 张设计图、施工图、效果图、规划图、超1500GB免费资源下载

第一建筑设计阅读互动平台

海量建筑项目全案 / 设计师、机构、供应商展示 / 真实身份认证 / 版权保护、便捷交易
微博、精选集、成长足迹，您最忠实的贴身管家
评论、分享、关注您心目中的大师

Tel：022-27403599 天津建邦数码科技有限公司
E-mail：architbang@yeah.net 977450152@qq.com

扫描二维码
微信关注建筑邦

International Landscape Wind Vane

国际景观风向标

李壮 编

图书在版编目（CIP）数据

国际景观风向标．2 / 李壮编．— 天津：天津大学出版社，2013.3

ISBN 978-7-5618-4621-6

Ⅰ．①国… Ⅱ．①李… Ⅲ．①景观设计 — 作品集 — 世界 — 现代 Ⅳ．① TU968.2

中国版本图书馆 CIP 数据核字（2013）第 031377 号

主 编：李壮

责任编辑：朱玉红

执行主编：石晓艳

编 委：李 秀 刘云 韦成刚 高松 赵睿

翻 译：国歌 樊喜强 王艳秋

编 辑：肖娟

设 计：北京吉典博图文化传播有限公司

出版发行：天津大学出版社

出版人：杨欢

地 址：天津市卫津路 92 号天津大学内（邮编：300072）

电 话：发行部 022-27403647 邮购部 022-27402742

网 址：publish. tju. edu. cn

印 刷：上海锦良印刷厂

经 销：全国各地新华书店

开 本：220mm×300mm

印 张：42

字 数：785 千

版 次：2013 年 4 月第 1 版

印 次：2013 年 4 月第 1 次

定 价：736.00 元（全两册）

EL:CH
008-011 Cantina di Terlan / Winery Terlan – Wine Garden
012-017 Landhauspark and Promenade
018-023 Urbanstrasse

GLASSER & DAGENBACH LANDSCAPE ARCHITECTS
026-029 A Star for Mies – Art Project for The 125 Anniversary of Architect Mies Van Der Rohe
030-035 Garten Von Ehren — An Exclusive Garden Marketplace in Hamburg
036-043 Berlin Moabit Prison Historical Park
044-047 A Garden as a Second Living Room for a Villa in Berlin Dahlem
048-053 Hotel Ifen — Five-Star Hotel in Austrian Alps
054-059 Redesign of Imchen Square and Waterfront Promenade
060-063 A Minimalistic Garden in a Forest in Vilnius Lithuania

CHARLES ANDERSON | ATELIER PS
066-069 Arthur Ross Terrace at the American Museum of Natural History
070-073 Bellevue Collection
074-077 Denver Arts and Crafts House
078-081 Hill House (Lil Big House)
082-087 Island Square

MCGREGOR COXALL
090-097 Ballast Point Park
098-101 Chang Gung Hospital
102-103 Former BP Site Public Parkland
104-107 Pimelea Play Grounds Western Sydney Parklands
108-111 The New Australian Gardens at the NGA Canberra

TOPOTEK I GESELLSCHAFT VON LANDSCHAFTS ARCHITEKTEN MBH
114-117 Friedrich-Ebert Square, Heidelberg
118-121 National Garden Show Schwerin 2009
122-125 Augsburg Wollmarthof
126-129 Theresienhöhe Railway Cover Munich
130-133 Sports Facility Heerenschuerli
134-137 Bergannstrasse 71, Berlin
138-141 The Big Dig
142-145 Ehrenbreitstein Fortress
146-149 Hackesches Quartier, Berlin
150-153 Südkreuz Train Station, Berlin

ASPECT Studios
156-157 8a The Terrace, Birchgrove
158-159 717 Bourke Street
160-163 Foley Park
164-167 Melbourne Convention Exhibition Centre
168-169 Shibagaki and Ng Residence
170-175 Sydney Olympic Park Jacaranda Square Landscape Design
176-179 Sydney Rhodes Lot 8 Tandara – Little Space, Big Courtyard
180-183 The National Emergency Services Memorial
184-189 Sydney Wetland 5 ESD + Landscape Design
190-193 Bondi to Bronte Coast Walk Extension Design Statement
194-197 Frankston Foreshore Precinct

RAINER SCHMIDT LANDSCAPE ARCHITECTS
200-203 Villa H. St. Gilgen
204-207 Villa Garden Bogenhausen

目录

CONTENTS

	208-211	Weser Quartier Bremen
ANDREA COCHRAN LANDSCAPE ARCHITECTURE	214-215	Peninsula Residence
ITEWORKS STUDIO	218-221	Manassas Park Elementary School Landscape
THOMAS BALSLEY ASSOCIATES	224-227	Gantry Plaza State Park
TERRAGRAM	230-233	Chronos
	234-237	Red Garden
HUGH RYAN LANDSCAPE DESIGN	240-241	Altar Ego (Show Garden)
	242-245	Baywatch
	246-249	Landfall
	250-253	Normandie
	254-255	Sequoia (Show Garden)
	256-259	Split Level
UMBERTO ANDOLFATO	262-265	Euroflora 2006
KEIKAN SEKKEI TOKYO CO., LTD.	270-275	Green Hills TSUYAMA
	276-279	Hotarumibashi Park
	280-283	Fukuoka—Zeki Sakura Park
	284-287	Shinagawa Chuo Park
CTOPOS DESIGN	290-293	West Seoul Lake Park
GUSTAFSON GUTHRIE NICHOL LTD.	296-299	Cultuurpark Westergasfabriek
	300-303	Old Market Square Nottingham
SAUNDERS ARKITEKTUR AS	306-309	Aurland Lookout
SIMONE AMANTIA SCUDERI	312-315	Roof Gardens
BNIM	319-323	City of Greensburg Main Street Streetscape
REED HILDERBRAND	326-329	Berkshire Boardwalk
NELSON BYRD WOLTZ LANDSCAPE	332-335	Carnegie Hill House

EL:CH

EL:CH Landscape Architects

El:CH was founded in 2005 in Munich, after severeal jointly won awards in various architectural competitions. Among them a first price in the international design competition für the Lower Austrian Garden Exhibition Tulln 2008 in 2003 and a first price for the redesign of Landhauspark and Promenade in Linz, Upper Austria. Since 2008 there is also a Berlin branch office in service.
Partners are Dipl. Ing. Elisabeth Lesche (TU Dresden, University of Guelph, Canada) and DI Christian Henke (University of Wien, TU Dresden).
El:CH's line of experience spans public space design up to larger scale urbanism. A major part of our projects is recruted by sucessful participation in international design competitions, also in cooperation with various architectural firms. It is our main goal to achieve a uniquely crafted solution for any given situation, suiting firmly within the ITES specific qualities. Intense work with plants and materials and thorough consideration of details are among our top priorities during the implementation of our designs.
The range of our projects covers private and public housing projects, urban parks and pedestrian areas as well as spaces directly related to or incorporated in buildings of cultural and other specific usage. Moreover we are involved in a number of conversational projects reclaiming former military spaces and urban planning.

Cantina di Terlan / Winery Terlan – Wine Garden

di Terlano 葡萄酒酒窖 / Terlan 葡萄酒厂—葡萄酒花园

LOCATION: Adige, Italy
项目地点：意大利 阿迪杰

AREA: 300 m^2
面积：300 平方米

COMPLETION DATE: 2009
完成时间：2009 年

PHOTOGRAPHER: Christian Henke, Elisabeth Lesche
摄影师：Christian Henke, Elisabeth Lesche

DESIGN COMPANY: EL:CH Landscape Architects
设计公司：EL:CH Landscape Architects

Cantina di Terlan / Winery Terlan – Wine Garden

With the extension of the traditional Cantina di Terlano, one of the oldest wine cellars of South Tyrol, a roof garden with a scenic view was added on top of the new outbuilding.

The wine garden is entirely dedicated to the experience of the spectacular landscape and thus closely connected to the locally produced wines. With its glass balustrades the garden merges seamlessly into the surrounding landscape.

di Terlano是南Tyrol地区最古老的传统葡萄酒酒窖之一，随着它的扩建，带着美景的屋顶花园建在外楼的顶部。

这个葡萄酒花园完全是为了让人体验壮观的景观而建的，因此它与当地产的葡萄酒息息相关。带着玻璃的围栏、花园与周围的景观紧密相连。

A sequence of green carpets and wooden terraces inspired by local field structures are arranged on a surface covered in porphyry gravel. Together with the change in material one low step marks the border between moving and resting areas.

受到当地田地结构的启发，在铺着斑岩砾石的平面上排布着连续的绿色地毯和木头平台。向下的一级台阶是活动区和休息区的边界，两个区的建筑材料不同。

The plantation combines a low, carpet-like plain with stripes of perennial grasses. Seasonally emerging bulbs, such as Allium sphaerocephalon, add transient colour. Fragrant herbs and a pomegranate tree planted in a square metal container complete the sensory experience, linking the garden to agricultural tradition.

Locally quarried porphyry, the warm colour of corten weathering steel and the lush green carpet combines to form a harmonious composition within a scenic setting.

植物区是一块低平地，像地毯一样，上面铺着条纹图案的多年生草坪。一些诸如圆头大花葱的花草随着季节露出球茎，增添了季节的色彩。

芳香的草本植物和正方形金属容器内的石榴树使感官的体验更加完整，使花园和农业的传统联系在一起。

当地挖掘的斑石，暖色的耐腐蚀高强度钢以及郁郁葱葱的绿色地毯结合在一起，在美丽的环境里构成了一副和谐的画面。

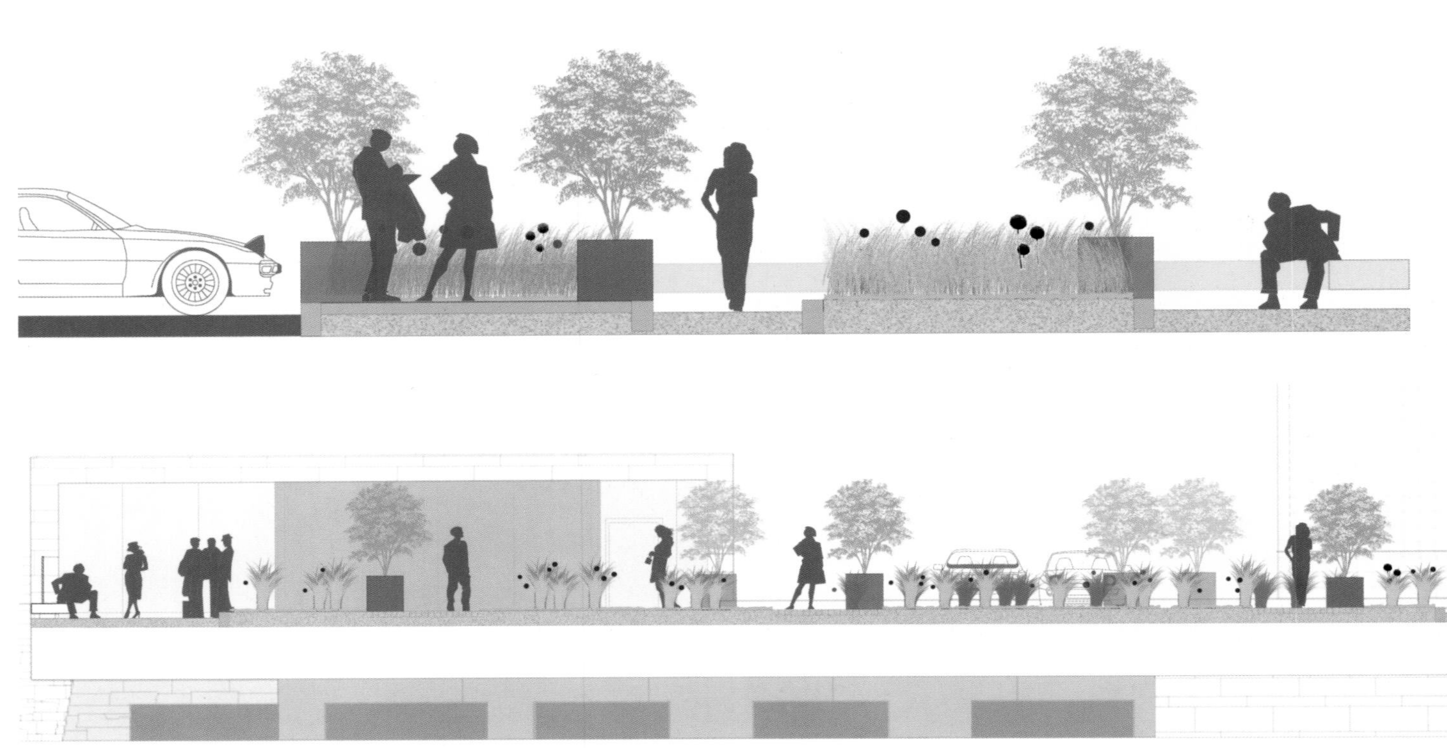

Landhauspark and Promenade

乡村别墅公园和散步广场

LOCATION：Linz，Austria
项目地点：奥地利 林茨

AREA：20,000 m^2
面积：20 000 平方米

COMPLETION DATE：2009
完成时间：2009 年

PHOTOGRAPHER：Christian Henke，Alexander Henke
摄影师：Christian Henke，Alexander Henke

AWARD：Sterreichischer Bauherrenpreis 2009(Austrian Client's Price)
奖项：Sterreichischer Bauherrenpreis 2009(Austrian Client's Price)

AWARD DATE：06．11．2009
获奖时间：2009 年 11 月 06 日

DESIGN COMPANY：EL:CH Landscape Architects
设计公司：EL:CH Landscape Architects

Landhauspark and Promenade

The site in its present form came into existence only in 1800 when a fire destroyed most of the city. The rampart was razed and transformed into a horticultural designed area. The 20th century's developments in transport had a negative impact on the space, making it primarily a passageway for motorized traffic. Pedestrian movement was limited to narrow areas, close to the buildings on both sides, and available space devoured by parked vehicles and an unimpeded clutter of urban furniture.

In 2005, the government of Upper Austria and the Linz City Council agreed in a joint effort to install an underground car park and remodel the area's surface. The resulting competition brief required entrants to develop a discernible identity for the site, providing usable space for a variety of user groups, while retaining all the existing trees, which are scattered without recognizable pattern all over the area.

After our preliminary site visit, we soon realized: the space's identity was already there. It had just become unrecognizable beneath layers of confusing additions and competing requirements. We observed a sophisticated scenic urban facade along the southern and western side of the site. The northern and eastern edges of the L-shaped plot were adorned with an amazingly pictorial population of trees. This bipolar identity, oscillating between the urban side of the actual Promenade and its green counterpart—the Landhauspark, was the atmospheric concept we strove to make comprehensible for the users.

Both sides draw from the other's complementing qualities: The Park gives a green vestibule to Promenade and Landhaus, connecting with the clos-by Castle Mountain. The Promenade stretches its "urban parquet" towards the city's main street along its unique architectural backdrop.

项目建筑场址的外观布局在1800年才形成，当时整个城市的大部分被一场大火毁掉。城池的围墙被夷为平地，然后在这块地上进行了园艺设计。20世纪交通的发展对该空间造成了负面影响，它基本上变成了一条机动车通道。步行只限于紧挨楼道的狭窄通道，可用空间被停放的车辆和一堆杂乱的家具吞没。

2005年，上奥地利州政府和林茨市委员会同意共同努力修建一个地下停车场，对整个地块的外部进行重新改造。随之而来的设计竞赛纲要要求参赛者为该场地打造出一种鲜明的特色，给各类用户群体提供可用的空间，但对散布在整个园区的无序排列的树要予以保留。

初步实地考察之后，我们很快意识到：这个空间的特征就已经在那儿了，只是在一层层覆盖的附加物和比赛要求的影响下变得模糊了。在这块L形场地的南边和西边，我们发现了优美的都市景观，在北侧和东侧边缘栽种着许多如画的树。一边是散步广场的都市气息，一边是绿色公园的乡村格调，这种双重风格正是我们努力向用户们所传达的意境。

两边的特色互为补充：公园与附近的城堡山相连，给散步广场和乡村别墅提供了绿色入口；散步广场像一块铺设在城市上的地板向城市的主街延伸，街边是独一无二的建筑风景。

Legend:
paved surfaces
granit pavements
compacted surface
syenite bench
vegetation surfaces
lawn
perennials
trees
internals
entry underground parking
wooden deck
Stadtparkett
grünes Vestibül
Promenade No. 1-31 -
Stadtparkett / Gehsteig
Lichtstelen
(Ladezone)
Fahrbahn
Gehsteig
grünes Band
Flanierzone
Steinband
Steinrahmen
Rasen-Diwan
Gartenweg
Vorgarten
Landhaus

Urbanstrasse

城市街区花园

LOCATION: Munich, Germany
项目地点：德国 慕尼黑

AREA: 2,000 m^2
面积：2 000 平方米

COMPLETION DATE: 2009
完成时间：2009 年

PHOTOGRAPHER: Christian Henke, Elisabeth Lesche, Michael Heinrich, Frank Sauer
摄影师：Christian Henke, Elisabeth Lesche, Michael Heinrich, Frank Sauer

AWARD: "Jung, schön und noch zu haben – die besten Immobilien Münchens" ("Young, beautiful and available" – Munich's Best Real Estate")
奖项："Jung, schön und noch zu haben – die besten Immobilien Münchens" ("Young, beautiful and available" – Munich's Best Real Estate")

AWARD DATE: 18. 11. 2009
获奖时间：2009 年 11 月 18 日

DESIGN COMPANY: EL:CH Landscape Architects
设计公司：EL:CH Landscape Architects

Urbanstrasse

The differentiated vegetation and simple, elegant materials create a surprising oasis in a densely and inhomogeneously built city block.

各色不同的植物和简单优雅的材料在建筑风格各异且密集的城市街区里营造出一个令人惊叹的绿洲。

A varied choice of housing forms for families and other user groups is fitted tightly within a limited urban space. Therefore, the finely balanced equilibrium between private and communal spaces is crucial. Areas of different privacies flow into each other without hard barriers, giving the space an atmosphere of Mediterranean ease. The town houses' privately used wooden decks to seperate a sheltered part behind wooden screens and an open part toward the brick wall along the property line. Users control the degree of interaction with their neighbours by moving the terrace furniture.

家用及其他用途的建筑形式各色各样，紧密地分布在一个有限的都市空间里。因此，私人和公共空间的良好平衡至关重要。不同的私人空间彼此衔接流畅，没有建筑障碍，给整个空间一种在地中海度假般的氛围。这些私人房屋使用的木头平台把木制屏风后的遮蔽部分和面向建筑边界砖墙的开放空间分隔开。使用者通过挪动平台上的家具来掌控与邻里的交流程度。

The neighbourhood's kids are using the connecting path between courtyard and playground as an additional space for play. It marks at the same time a buffer space towards the neighbouring property. Wooden seating boxes create a protective distance without disrupting the view to the court. The open space layout has proved itself in practice, helping to create a harmonious neighbourship quickly.

街坊四邻的孩子把院子与操场间的连接小路作为额外的玩耍空间。与此同时，它成为了通向附近建筑的缓冲区。休息用的小木头亭子在距离上保证不会干扰人们欣赏院子的景色。这个开放空间的布局实际上已经证明了它有助于很快建立起一种和谐的邻里关系。

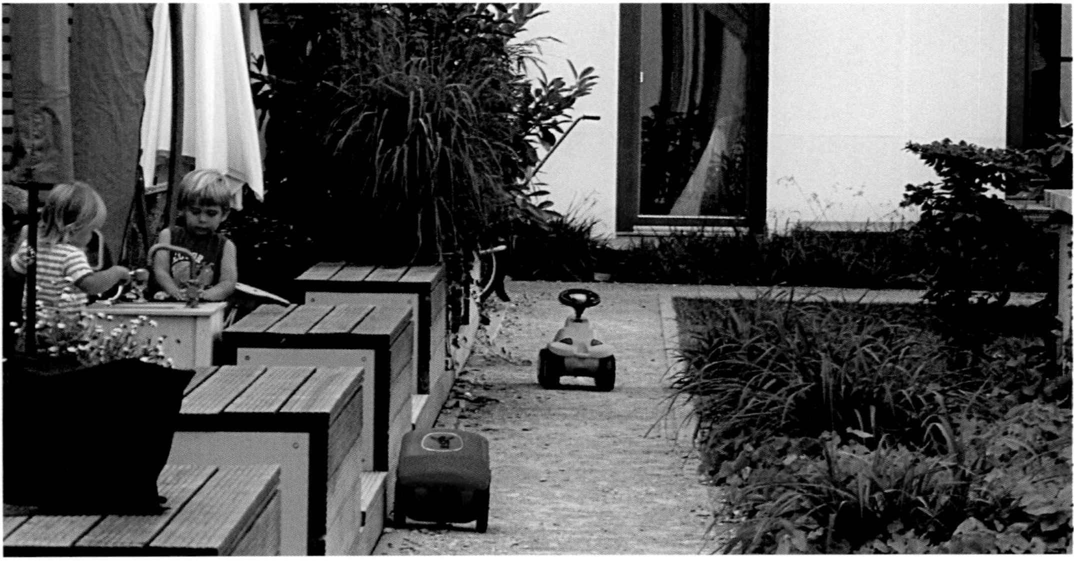

Minimal variety of materials for maximal conciseness:
Horizontally and vertically oriented wooden surfaces repeat throughout the space. Their repetition gives a common identity to the nooks and crannies of the site. Moreover they imply a connection with the site's former use as a carpentry's storage yard. The strong framework gives a letter-case-like base for the individual appropriation of space.
A special place is created with a three-part wooden deck converging around a single tree.
The multi-use meeting area marks the neighbourhood's heart.
Pathways inside the court have loosely gravelled and compacted surfaces.
Sidewalks in front of the building blend seamlessly into the environment by employing commonly used Munich sidewalk flagstones.

最少种类的材料达到最大程度的简洁效果。
在整个空间里，水平和垂直方向的木制平台不断出现。这种重复给该空间的每个角落赋予共同的特征。而且，它们还隐约地体现出该场地的前身是一个木料房。鲜明的结构框架像文件夹一样为个人的空间做分配。
一个特殊的空间被打造出来，带有一个三部分组成的木头平台，围绕在单棵树周围。多功能会议区标志出这是街坊四邻的中心。
院子里的路由松散的卵石铺成，很坚固。楼前的人行道铺了常用的慕尼黑步行石板，与周围环境紧密融合。

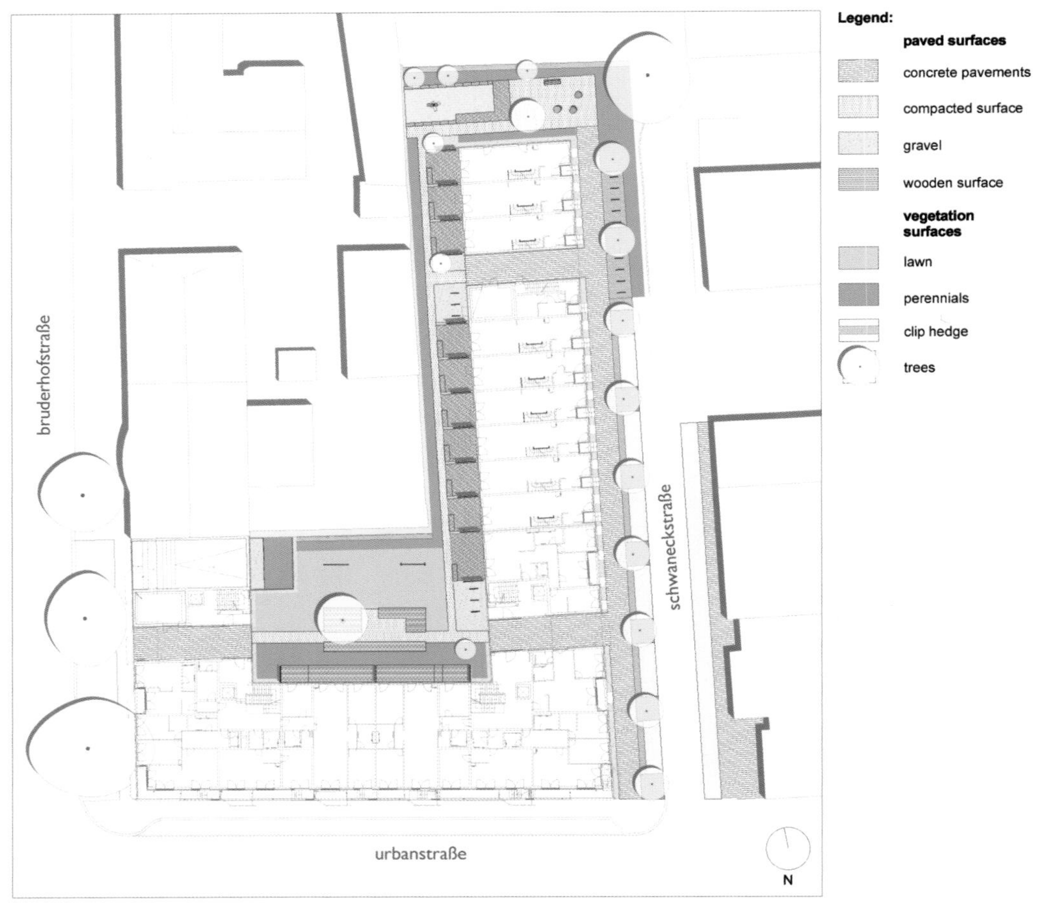

The playground, located in the northern corner of the site, goes with the general design principle, comprising a wooden deck and a flowering perennial frame. The several-storey firewall at the playground's front end is used for ball games and radiates warm colours.
A single tree creates a special focus inside the yard. The property's edges are framed with lush greenery, not only employing hedges and climbing plants, but mainly perennials. Plantation combines specifically chosen, shade-loving perennials. Sturdy ground covers, as Luzula sylvatica and Bergenia cordifolia come together with flowering specimen as Astilbe chinensis var pumila and Gillenia trifoliate to create a seasonal hanging carpet.

操场位于项目地址的北角，符合总的设计原则，包括一个木头平台和一个多年生花卉架。位于操场前端几层楼高的防火墙可服务于球类运动，散发着温暖的色彩。
院子里唯一的一棵树成了一个焦点。该区域的四边围着郁郁葱葱的植物，不仅有树篱，还有攀爬植物，但主要是多年生植物。种植园多采用喜阴多年生植物。地杨梅和岩白菜这些生命力顽强的地被植物和一些矮生落新妇和美吐根花交织在一起，构成了随季节而变的“挂毯”。

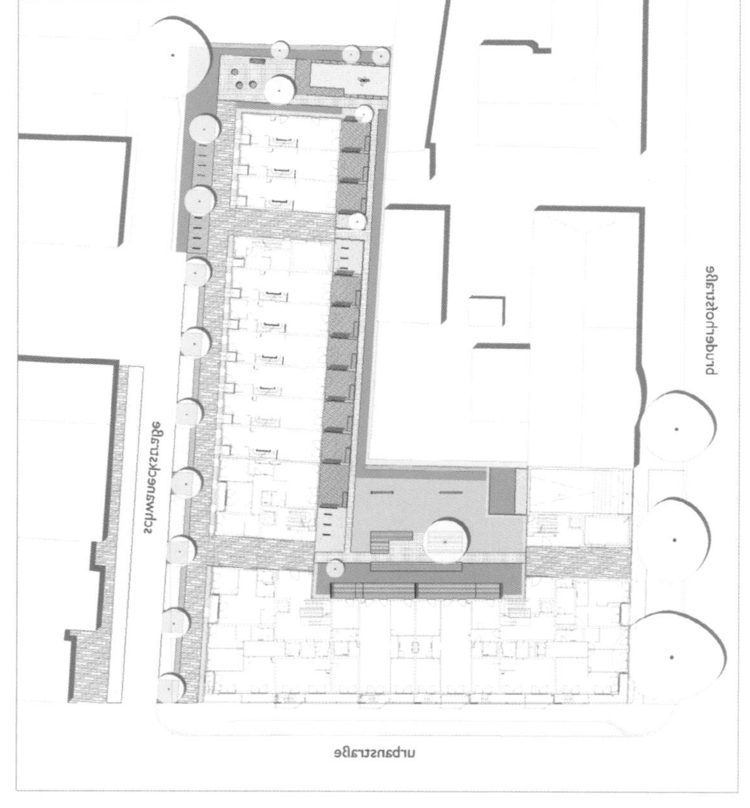

DAS DIESEN
BAU ERFÜLLT,
MAUER
RK UND

Clients:

Clients are both public and private as for example:
Travel Charme Hotels & Resorts AG, Zürich (CH)
City Hall of Berlin,Germany
City Hall of Batumi ,Georgia –Republic of Adjara
Gazprom (Resorts Devision)
Aspria Holding NL London (Wellness and spa clubs)
miesvanderrohehaus gallery Berlin ,Germany
Development solutions Moscow, Russia (businessparks)

Awards:

German landscape architecture award bdla – 2007 1st prize
Austrian daylight spaces international design and architecture award 2007 – 1st prize
Made in Germany – chapter landscape – Braun publishers ,Hamburg , Germany 2007 – 2nd prize

Philosophy:Gardens and Parks are Backdrops

The settings of gardens and parks form backdrops before which visitors, whether public or private, are able to act out a role in their very own play.
The manner in which we approach the design of gardens, parks and landscapes depends entirely on the character of the space and how it will ultimately be used.
We consider both the shape and structure of the existing surroundings and, naturally, the needs and wants of the visitors or clients, and then act as an intermediary and instigator between the space and the user.
We see ourselves as a tool which can be used to lend shape and expression to the conscious and subconscious wishes of clients or visitors.
The design must be strong and clear so as to bestow lasting energy, expression and purpose upon the open space we have crafted.
Our aim is to create gardens and parks with which we can identify on an emotional level, yet which still retain a lasting, timeless clarity.
Our garden-creations are therefore not restricted by genre. We are able to bring our ideas to fruition whatever the desired style is.
A contemporary style is neither a prerequisite nor a hindrance to achieving high-quality design.

Mr. Udo Dagenbach
University Diploma (Landscape Architect), Technical University (Berlin) 1986
State approved landscape gardener
Stone sculpting since 1994
guest student at the University of Art, Berlin at Makoto Fujiwara s stone sculpting class
1985—1987 Colleague of the Japanese sculptor Professor Makoto Fujiwara main project sculpture and garden at the Bundesanstalt für Geowissenschaften Hannover — land art project

Mrs. Silvia Glaßer
University Diploma (Landscape Architect). University of Nuertingen 1985
State approved Gardner specialized in perennials

GLASSER & DAGENBACH LANDSCAPE ARCHITECTS

GLASSER & DAGENBACH LANDSCAPE ARCHITECTS

Partnership dedicated to high quality Landscape and Garden Architecture.
Colleagues Silvia. Glaßer and Udo. Dagenbach formed the Partnership during 1988.
Office team:
mostly 3 landscape architects, 2 drafts women, 2 -3 technical assistants

The office is engaged mainly in new construction of public parks and private hotel, wellness and Resort projects. Reconstruction of listed gardens and parks are very welcome projects too.
Beside regular bread and butter work the team likes to cross the border to land art and sculpture.

密斯之星——纪念密斯·凡德罗 125 周年的艺术工程

LOCATION：Berlin，Germany
项目地点：德国 柏林

AREA：300 m^2
面积：300 平方米

COMPLETION DATE：2011
完成时间：2011 年

PHOTOGRAPHER：Udo Dagenbach
摄影师：Udo Dagenbach

DESIGN TEAM：Sabrina Schr der，Marina Kanzler
设计团队：Sabrina Schr der，Marina Kanzler

LANDSCAPE ARCHITECT：Udo Dagenbach
景观设计师：Udo Dagenbach

DESIGN COMPANY：Glasser & Dagenbach Landscape Architects
设计公司：Glasser & Dagenbach Landscape Architects

A Star for Mies – Art Project for The 125 Anniversary of Architect Mies Van Der Rohe

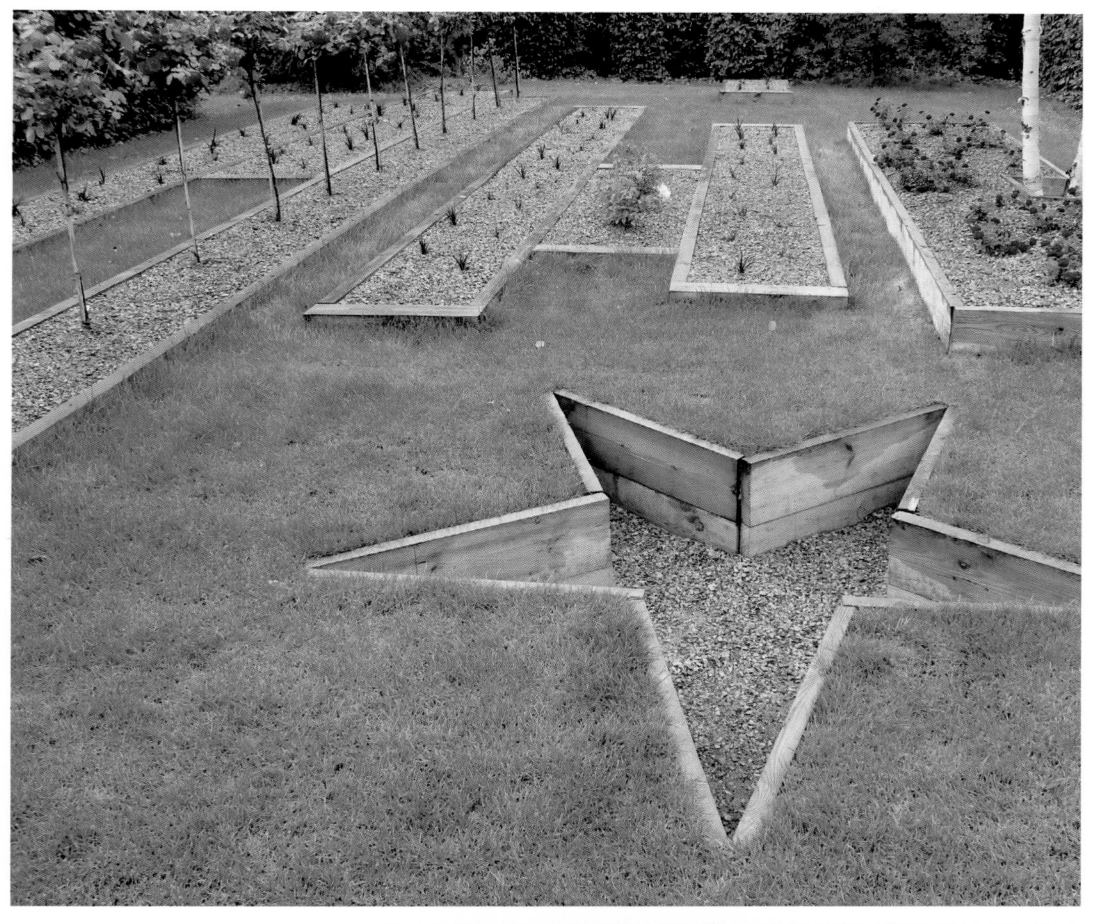

Ludwig Mies van der Rohe designed the 1926 built revolution monument dedicated to the murdered socialists Rosa Luxemburg and Karl Liebknecht. It existed until its destruction through the Nazis in 1935 on the cemetery in Berlin Friedrichsfelde. It was an outstanding piece of art with enormous international recognition and far ahead of the architecture at that time.

由密斯·凡德罗设计、1926年建造的革命纪念碑，目的是纪念两位被害的社会主义者——Rosa Luxemburg和Karl Liebknecht。它只存在了短短9年，于1935年在柏林的弗雷德里司福德被纳粹摧毁。它在当时是一件艺术杰作，设计理念超前，得到国际社会的广泛认可。

The so-called Landhaus Lemke, a very modern and modest bangalow in Berlin Hohenschoenhausen, located at the so-called Obersee Lake, was the last work of Mies in Berlin before he emigrated to America.

所谓的蓝德豪斯·勒穆克，是柏林Hohenschoenhausen地区的一座非常现代朴素的小房子，它位于Obersee湖畔，是设计师密斯·凡德罗移居美国之前在柏林的最后作品。

The land art project uses design principles of the destroyed monument. The process of the destruction is found by deconstructing the monument elements and it is reconstructed in a changed symbolic meaning.

这项艺术工程采用被纳粹摧毁的纪念碑的设计理念。通过解构纪念碑的元素，设计者发现了整个破坏过程，然后以一种经过改变的象征意义重建这座建筑。

The shifted and stacked brick stone cubes have changed to parallel plant beds framed by wooden blankets.
They are planted with red-leafed grass (Ophiopogon planiscapos nigrescens), red flowering high stem roses and ground cover roses and white park roses which Mies liked very much.

错综堆积的砖和石块已被重整，与四周用木板围起的花坛平行。
花坛里面种上红叶草（一种麦冬），红色的高茎玫瑰和地被玫瑰以及白色的公园玫瑰，这些都是密斯喜欢的植物。

The plant beds are mulched with crushed brick stones. This puts a strong contrast to the lawn.

花坛用碎砖石做护根层，这和柔软的草坪形成强烈对比。

The Soviet star as an element of the former monument is stamped about 40 cm into the ground and forms a new communication point like a Japanese kotatsu table.

原纪念碑上的象征苏维埃的五角星被嵌入土里 40 cm，形成一个新的交流点，形状就像日本的暖桌。

The ground of the star is filled with crushed brick stones. This way the star gets an additional function beside its historical meaning and transforms to a sign as in the design sketch of Mies.

五角星下面的土地也填上碎砖石，这样，在它原来的意义之外，又有了新的意义，它已经变成一个符号，就像密斯在他的设计草图上常用的符号。

An edition of 10 special-shaped brick stones uses elements of the revolution monument and helps to find the design idea of the land art project.

这种设计使用了 10 种特殊形状的砖石，表现了革命纪念碑的艺术元素，这会帮助人们理解这项地上艺术项目的设计理念。

Garten Von Ehren—An Exclusive Garden Marketplace in Hamburg

von Ehren 花园——一个独特的汉堡花卉市场

LOCATION：Hamburg，Germany
项目地点：德国 汉堡

AREA：13,000 m²
面积：13 000 平方米

COMPLETION DATE：2007
完成时间：2007 年

PHOTOGRAPHER：Udo Dagenbach，Arnt Hauk（central perspective）
摄影师：Udo Dagenbach，Arnt Hauk（central perspective）

DESIGNERS：Udo Dagenbach，Silvia Glasser
设计师：Udo Dagenbach，Silvia Glasser

DESIGN COMPANY：Glasser & Dagenbach Landscape Architects
设计公司：Glasser & Dagenbach Landscape Architects

Garten Von Ehren — An Exclusive Garden Marketplace in Hamburg

The Lorenz von Ehren Nursery—founded in 1865—is a very traditional German Nursery situated in Hamburg and Bad Zwischenahn. The nursery is worldwide well-known for its special-shaped big trees. Our office has worked together with them since more than 15 years ago.

The owner family situated in Hamburg wanted to do a relaunch of the garden centre adjusted to headquarters of the company.

洛仑兹·冯·艾伦苗圃始建于1865年，这是一个非常传统的德国苗圃，位于汉堡和巴特茨维什安。该园以其形态各异的大树而享誉世界。我公司与其业务往来已经超过15年。

苗圃主人一家住在汉堡，他们想要重新修整中心花园，使其适应公司总部的风格。

The family decided in 2006 to build a new glasshouse with an almost 2 hm^2. sales garden around it and to demolish the old garden centre.

The task for our office was to create both a relatively static, representative entrance situation and a flexible sales garden with a upgrade and stylish look. A lot of existing materials and plants had to be integrated. During the realization many new materials of sponsors had to be integrated respectfully and expressively.

All this produced by itself a process of continuously renewed and changed hand drawings. CAD drawings were really the last step and changed very often, too. A very unusual theme needed a very unusual approach to the design.

2006年，家族决定重新建立一座暖房，在它周围修建一个达到2公顷的花卉销售园区，并将原来的中心花园拆除。

我公司的任务就是建立一个相对静态的且有代表性的花园入口，还有一个高档、漂亮且灵活的花卉销售中心。设计时要求许多现有的原料和植物必须相互融合。在实现这一设计的过程中，赞助商的许多新的材料必须加入其中，要让人感到舒适并且具有表现力。

所有的这些需要不断地更新改变手绘图。CAD图纸确实是最后一步，但也经常有改动。不寻常的主题需要不寻常的设计手法。

In front of the entrance I had the idea of a kind of Broderie in a mixed modern historical style, tribute to the tradition of the nursery and the need of showing new elements. A circle of 9 m formed a gem-like plant bed in which about 4,000 buxus were used. The picture which the photographer Arnt Haug took was taken two days after the opening of the garden centre and three days after planting the buxus. Only the best plants of the nursery were used in this part.

在入口的前端，设计一个融入历史与现代风格的刺绣作品，表现对苗圃悠久传统的敬意，同时也展现了新元素。一个为9米的圆构成了一个像宝石一样的花坛，里面栽有4 000棵黄杨。这张照片是摄影师阿尔恩特在中心花园开业两天后和栽种完黄杨三天后拍摄的。只有苗圃中最好的植物才能在这儿使用。

In front a very classical French style ornament was used in the inverse way—the ornament was white gravel and the rest was buxus. To the right circle-and-ball-shaped forms of buxus sempervirens, Taxus baccata and Ilex crenata created a playful and almost amorphe garden. There the paths were amorphe forms with black copper cinder chippings. To the left the idea of a formal garden in constructivist style was developed from an earlier garden art project in the Berlin Mies van der Rohe Haus. Flat and high cube-formed Taxus baccata, and buxus sempervirens were contrasted with umbrella-shaped Pinus sylvestris.

在前面，采用逆向的方式用了一个很具有古典法式风格的装饰——装饰物是白色砾石，而其他部分是黄杨。在右边，圆形球状的黄杨、杉树、红豆杉和钝齿冬青木创造了一个有趣的、随性的苗圃。那儿的小路也是随性的，上覆黑色的铜渣碎屑。在左边，受到在柏林Mies van der Rohe Haus的一个早期园艺项目的启发，意图构建一个具有构成主义风格的正式花园。平坦、高大、立方体形的红豆杉和黄杨、杉树与伞形的樟子松形成鲜明对比。

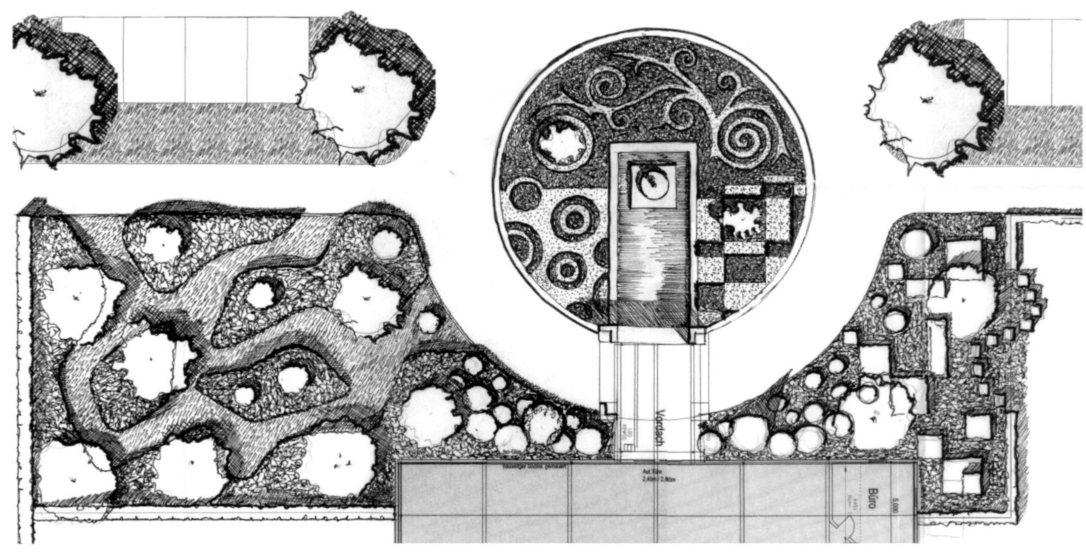

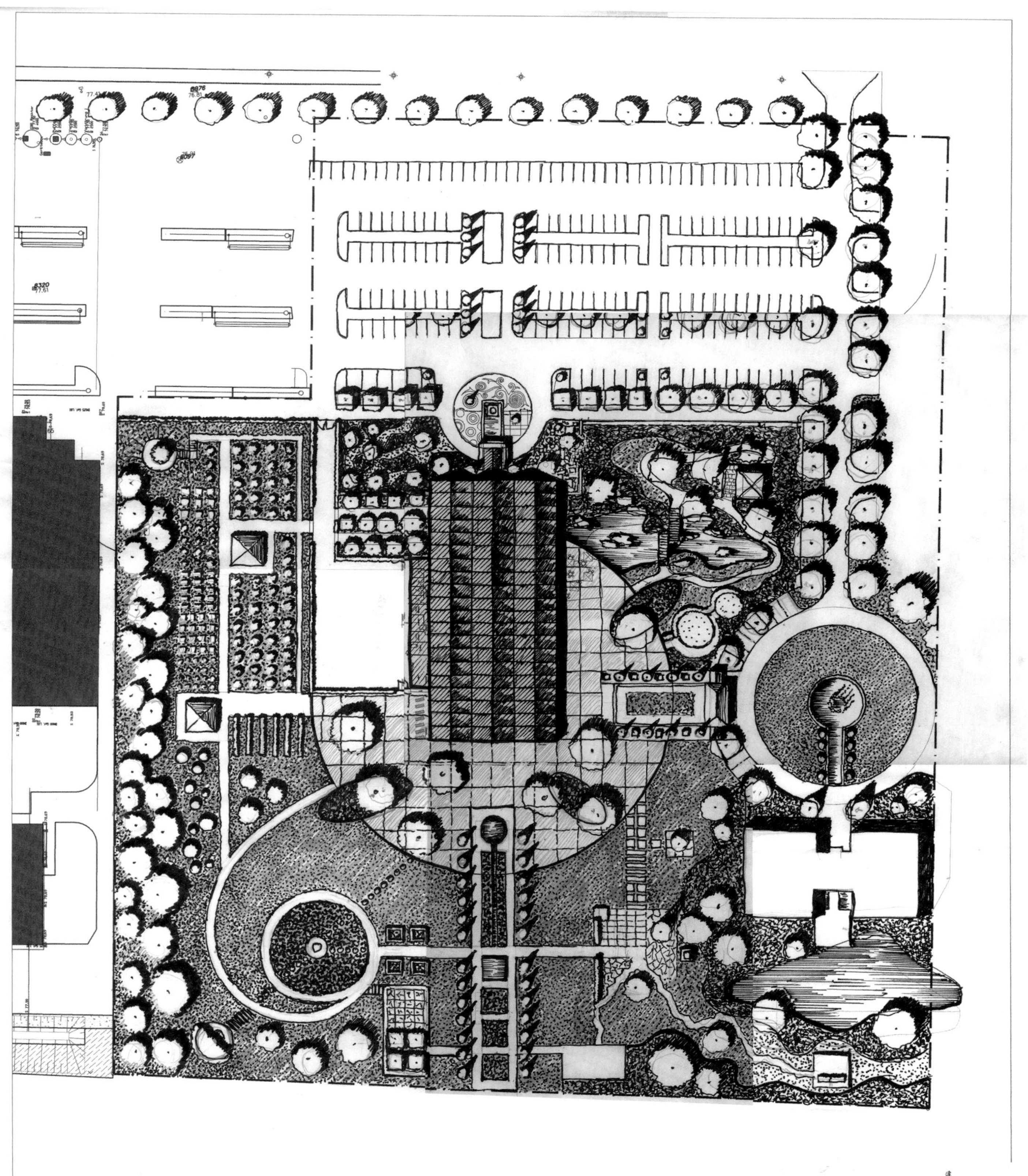

Berlin Moabit Prison Historical Park

柏林莫阿比特监狱历史公园

LOCATION：Berlin，Germany
项目地点：德国 柏林

AREA：30,000 m^2
面积：30 000 平方米

COMPLETION DATE：2006
完成时间：2006 年

LANDSCAPE ARCHITECT：Udo Dagenbach，Glasser and Dagenbach
景观设计师：Udo Dagenbach，Glasser and Dagenbach

PHOTOGRAPHER：Udo Dagenbach
摄影师：Udo Dagenbach

COLLABORATOR/ASSOCIATE ARCHITECT：Alexander Khomiakov — drawings，Andreas Steiner，Leighton Pace，Martina Levin，Sabrina Schr der，Sabine Linke，Doerte Schroerschwarz，Katrin Weinke
合作者／助理建筑师：Alexander Khomiakov — drawings，Andreas Steiner，Leighton Pace，Martina Levin，Sabrina Schr der，Sabine Linke，Doerte Schroerschwarz，Katrin Weinke

Berlin Moabit Prison Historical Park

DESIGN COMPANY：
Glasser & Dagenbach Landscape Architects
设计公司：Glasser & Dagenbach Landscape Architects

The park's theme and urban planning and its architectural and political history are unique to Berlin's urban landscape. The task of creating both a memorial and an area for people to relax and learn has been accomplished in an exemplary manner. Historical landmarks have been preserved, restored and enhanced by using contemporary style.

这座历史公园的主题、城市规划以及它的建筑和政治历史对于柏林的城市景观来说是独一无二的。建设一片既纪念历史又适合人们放松和学习的地方已经以一种示范性的方式实现。历史遗迹得以保存、修复并用现代风格加强。

The dramaturgical approach of minimalist sculptural design principles once again anchors the structural remains permanently in the disordered urban space of the neighboring Central Station. Local people and foreign visitors to Berlin can rediscover the site's historical significance after more than 50 years of inaccessibility and enjoy its recreational resources.

极简主义雕塑设计原则中的拟剧场方法再一次永久性地把附近中央车站无序空间里的建筑遗迹保存下来。该历史遗迹已有50多年不对外开放，现在当地居民和到柏林的游客可以重新领略它的历史意义并享受它的娱乐休闲资源。

Local borough residents were deeply involved throughout the almost 16 years of planning and developing of the park. The Borough of Mitte, represented by the Streets and Parks Office, would like to submit this bid to illustrate how complex urban spaces can be upgraded and maintained through energetic, yet prudent landscaping.

当地自治区的居民全力参与该公园几乎16年的规划和开发。Mitte区以街道和公园办事处作为其代表提交投标书以说明如何通过积极但认真的景观建设来改造和维护复杂的城市空间。

The design for the history park is the result of an intensive study of the site's history, beginning with Moabit Prison construction 150 years ago. The park is replete with hints and references to the physical layout and the former use of the grounds.

该历史公园的设计建立在对该地区历史的深入研究之上，从150年前莫阿比特监狱的修建开始。该公园处处能让人感受到这些土地的自然布局和先前的功能用途。

The park is enclosed on three sides by the five-metre-high prison wall which remains intact. The wall and the three former guard dwellings (No.18) give visitors a good idea of the size and shape of the prison.

该公园三面由5米高的监狱墙围合，这些墙至今仍完好无损。这些墙和三个曾经的警卫住处(18号)使游客对监狱的大小和形状有一个很好的了解。

Visitors can enter the park through three variously designed entrances (No.1–3). Inside the park, sheltered by the high walls, the star-shaped layout of the former prison building is recreated. Wings BD (No.5–7) are depicted by sunken or elevated lawn levels. Hedgerows depict what was once Wing A and show the arrangement and size of the solitary cells (No.4). Visitors can explore a reconstructed cell in its original dimensions (No.4a) while listening to a recording of Albrecht Haushofer's "Moabit Sonnets" written during his incarceration in winter 1944—1945.

游客能够通过三个不同的入口（1~3号）进入公园。在高墙庇护下的公园内部，重新展现了以前监狱建筑呈星状的布局。公园的BD两翼(5~7号)是高低不平的草坪。灌木树篱则在曾经的A侧区，并展示了独立牢房（4号）的布局和大小。
游客能够参观一个按照原有尺寸（4a号）重建的牢房，并能听到Albrecht Haushofer在1944—1945年冬被监禁时写下的《莫阿比特十四行诗》的录音。

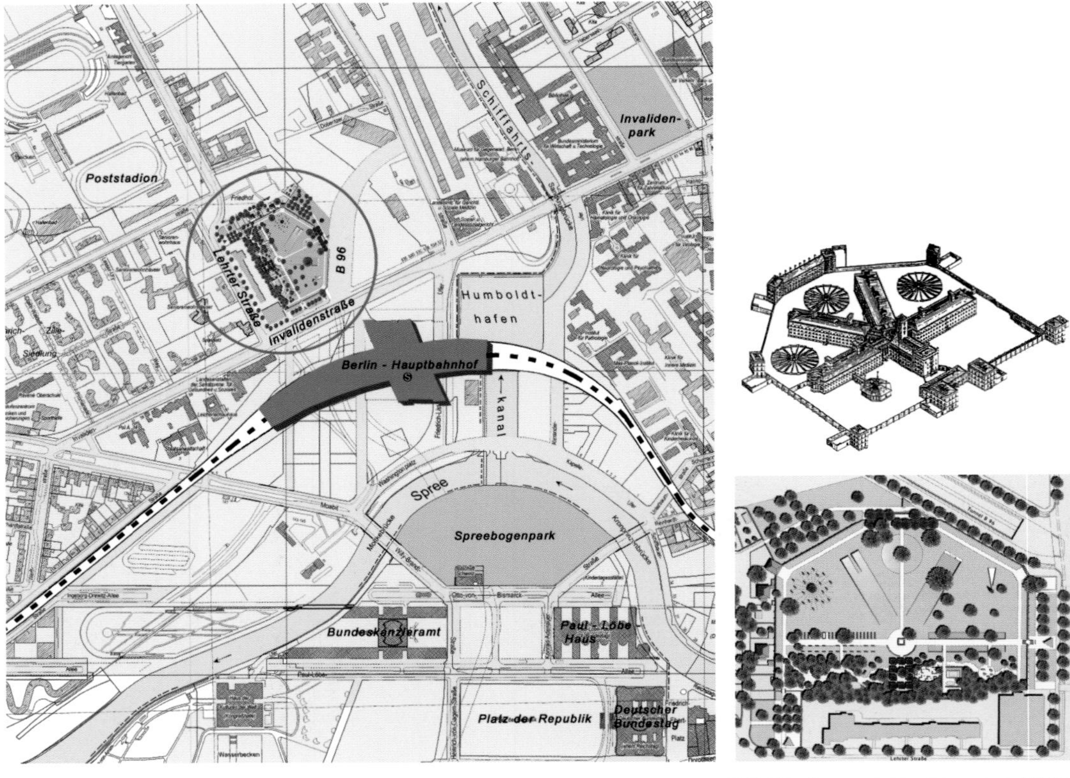

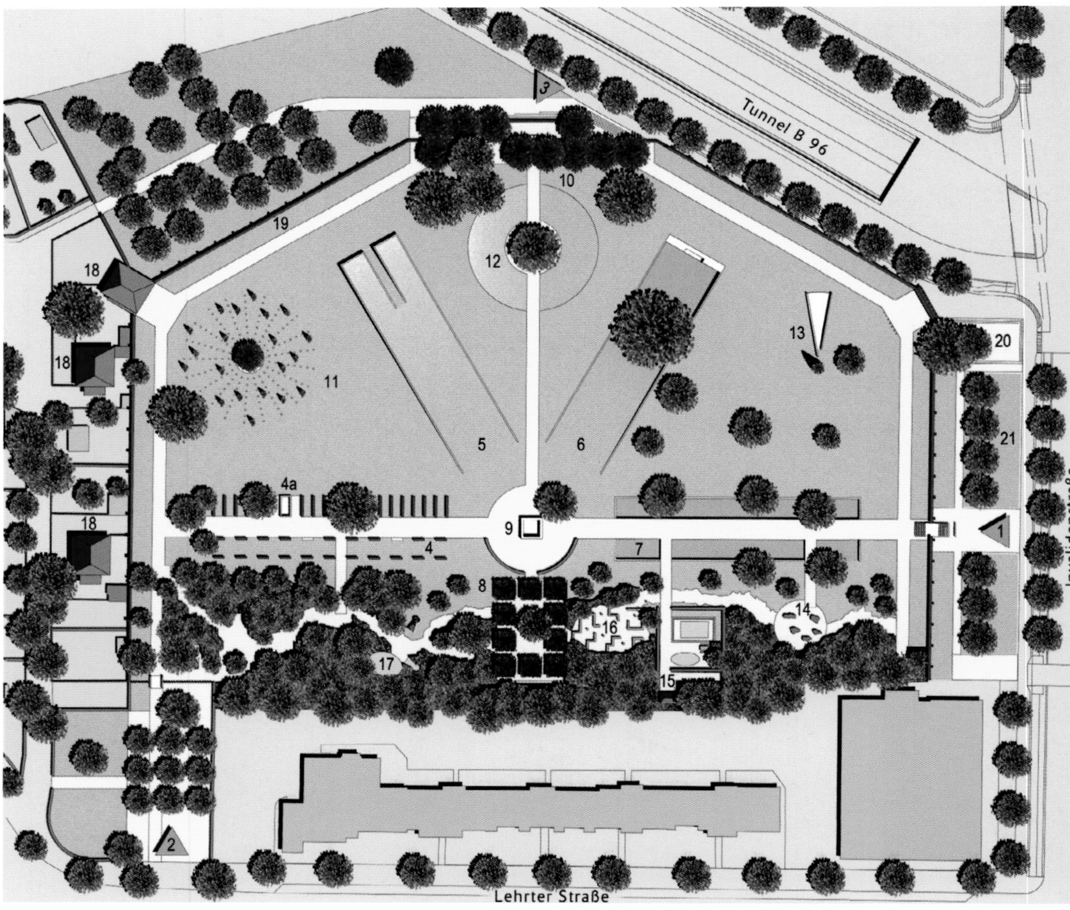

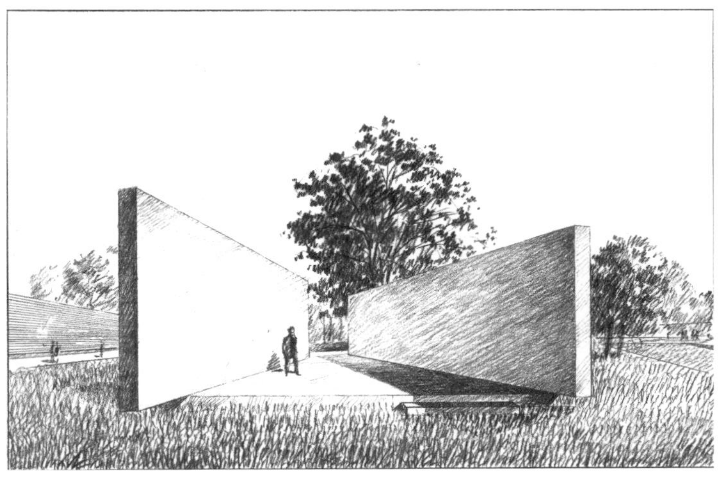

EIN HAUCH LEBENDIG, EINGEHEIMES ZITTERN...

DIESEN BAU
MAUERWERK UND EISENGITTERN EIN HAUCH LEBENDIG

A Garden As a Second Living Room for a Villa in Berlin Dahlem

柏林达勒姆别墅第二起居室花园

LOCATION: Berlin, Germany
项目地点：德国 柏林

AREA: 1,500 m^2
面积：1 500 平方米

DESIGN COMPANY: Glasser & Dagenbach Landscape Architects
设计公司：Glasser & Dagenbach Landscape Architects

A Garden As a Second Living Room for a Villa in Berlin Dahlem

We were asked to do a modern and reductive design which provided possibilities to install sculptures in the near future.

我们按要求做出一个既时髦又简约的设计，这样，在不久的将来，我们可能根据这个设计安装一些雕塑。

The views of neighbours should be screened by big trees and shrubs.
Existing very big oak trees had to be respected and in a hidden garden part a playground for the kids should be installed.

大树和灌木应该遮住附近的视野。现有的大橡树必须受到重视，而且还要在花园一个隐蔽之处建造一块可供小孩子们玩耍的游乐场。

Thus we decided to create a more spacious and structured garden which would give the owner and visitors a protected but yet very rich impression.
The trees of all neighbour garden were integrated by using them as second screen behind the first screen plantation.

因此，了解任务之后，我们打算建造一个更加宽敞、结构层次分明的花园，这样的设计会给花园主人和参观者安全感和富丽堂皇的印象。
所有附近花园的树都被融合进来，成为一种植园屏障后面的第二道屏障。

We changed the terrace area close to the living room to a spacious area with a beige natural paving of jurrasic marble from Bavaria.
To provide a strong screen we installed a long 2 m–high wall made of concrete and rendered with slabs of Jurrasic marble—same as the paving—with an integrated waterfall coming out of the wall and falling into a small basin.

我们把靠近起居室的阳台变得更加宽阔，并用巴伐利亚侏罗纪大理石铺设一条米黄色、显得很自然的小路。为了找一个结实的屏障，我们用混凝土和铺砌小路用的一样的侏罗纪大理石板建造了一堵长长的高达两米的墙。一个小瀑布从这座高墙上流下来，流入下面的小水池中。

Cube—shaped Carpinus betulus were used as 5 m—high screening plants .
Fagus sylvatica atropunicea and Fagus sylvatica were used as formal clipped 5m high screening plants Fagus blocks , cubes and hedges. They were completed with yew hedges and boxwood hedges. Japanese marbles and Amelanchier lamarckii which are both umbrella—shaped grow out of the formal clipped plants. Rhododendron, azaleas, cornus kousa and prunus cerasifera nigra are forming the counterpart to the formal plants. Decorative grasses, astilbe and hosta, some groundcover roses with lavender are the perennial layer in the setting.

立体状的欧洲鹅耳用做5米高的屏风植物。紫叶欧洲山毛榉和普通山毛榉修剪成整齐标准的长方体、立方体及树篱。它们是用紫衫木树篱和黄杨木树篱做成的。日本大理石和拉马克唐棣属都形似雨伞，从正式的修剪植物中长出。杜鹃花、四照花、紫叶樱桃李都和正式的植物相当。装饰性植物、落新妇属植物和玉簪属草本植物、一些地被植物、玫瑰和薰衣草，它们都是绿化设置上的多年生宿根层草本植物。

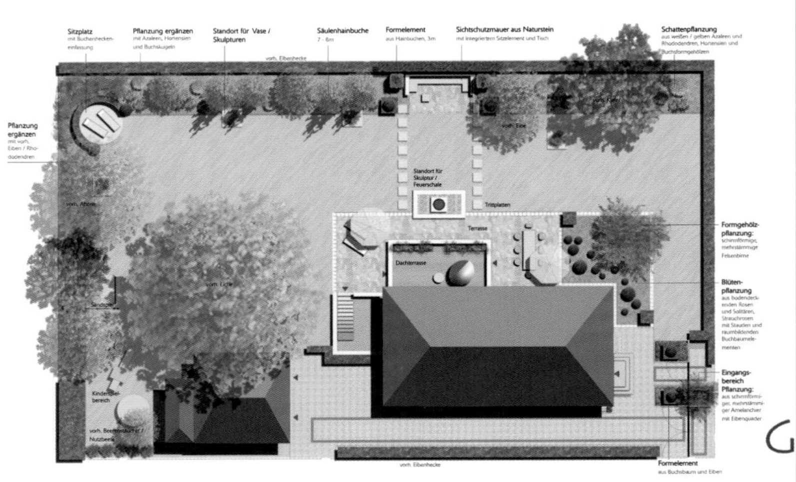

Entwurf Variante 1

The colours of the flowers are white and lila blue, Bulbs as daffodils, snow crocus and others follow the same colour principle.

花的颜色都是白色和淡紫蓝色。像水仙花一样的鳞茎植物、雪番红花、还有其他的花都遵循同一花色原则。

Entwurf Variante 2

Right now the new main entrance and the driveway are constructed by using the previously named Jurassic marble and black basalt paving.

现在，要用先前提到的侏罗纪大理石和黑色玄武岩石铺设新的主要入口和车道。

The whole planning and constructing became a process of very close discussion between client and planner. The trust the client gave to the planners to choose plants and materials supported the result very much.

整个项目计划和实施的过程就是客户和设计者密切讨论的过程。客户非常信任设计者，让设计者随意选择植物和建筑材料，这种信任有利于创造这样的作品。

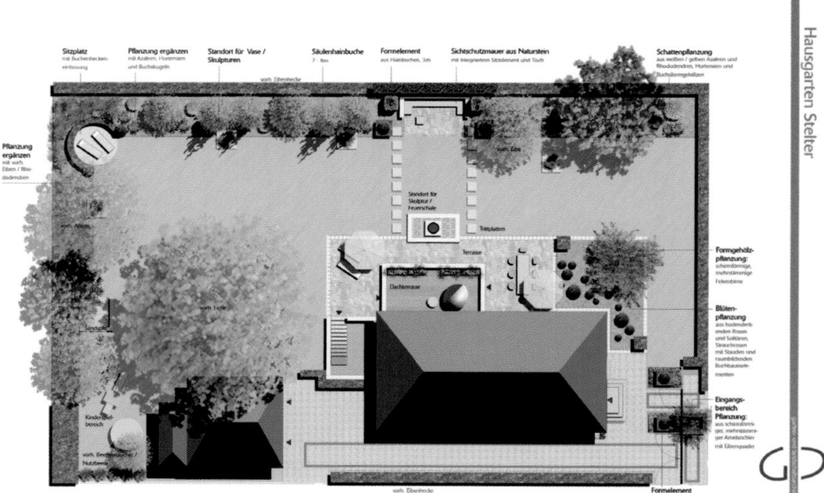

Entwurf Variante 1

Ifen 酒店——奥地利阿尔卑斯山五星级酒店

LOCATION：Kleinwalsertal，Austria
项目地点：奥地利 小瓦尔瑟塔尔

AREA：18,000 m^2
面积：18 000 平方米

COMPLETION DATE：2010
完成时间：2010 年

PHOTOGRAPHER：Udo Dagenbach
摄影师 ：Udo Dagenbach

DESIGNER：
Glasser and Dagenbach，Udo Dagenbach，Silvia Glasser，Sabrina Schroeder，Sabine Linke
设计师 ：
Glasser and Dagenbach，Udo Dagenbach，Silvia Glasser，Sabrina Schroeder，Sabine Linke

DESIGN COMPANY：Glasser & Dagenbach Landscape Architects
设计公司：Glasser & Dagenbach Landscape Architects

Hotel Ifen — Five-star Hotel in Austrian Alps

The Travel Charme Group decided to restart and refurbish a very traditional five-star hotel in a small beautifull valley located in Kleinwalsertal in the Austrian Alps.

在位于奥地利阿尔卑斯山脉的小瓦尔瑟塔尔一个美丽的小山谷中，Charme旅行集团打算重新建立和装修一个传统的五星级酒店。

The site is located on a steep ground with 14 m height differences on 1,100 m above sea level.

地块位于一个高于地面14米的陡峭斜坡上，海拔1 100米。

It is a mountainous region with mountains up to 3,000 m. It is a very traditional summer and winter tourism area.

这里是山区，山的高度为3 000米。因此，这里是一个传统的夏季避暑和冬季疗养的圣地。

Our job was to design the 1.9 hm^2. area according to the regional ecological situation, which means 2 m snow in wintertime, about 1,700 mm rain from springtime to autumntime and about −29°C in wintertime.

我们的工作是根据该地区生态环境设计面积为1.9 公顷的地块，这里冬季降雪量为2米，春天到秋天的降水量为1 700毫米，冬季气温大约零下29 摄氏度。

A big part of the site is so steep that normal planting or any use is not possible. To cut down costs we decided to install a grass landscape with mountain species which are adapted to the local conditions.

该地域的大部分地区是如此陡峭以至于不可能种植植被和被使用。为了降低费用，我们决定用适应该地区条件的山区物种设置草地景观。

For the shape of the site is very inhomogeneous, we had the idea to spread leaf-shaped flowerbeds all over the site

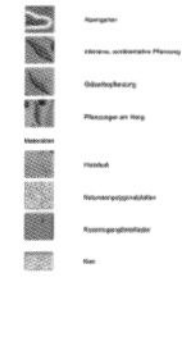

Neubau Ifen Hotel
Hirschegg

Entwurf
Aussenanlagen
Schauplan

like thrown by chance or blown there by winds.

As to the intensively used areas like terraces we covered with yellow Chinese granite slaps 15 cm x 35 cm and 50 cm x 50 cm.

由于该地区地形迥异，我们打算按照叶子的形状设计花坛，这就好像叶子偶然或随风飘落一样，把花坛四处散布开。像平台这样集中使用的地方，我们使用15 cm × 35 cm 和 50 cm × 50 cm 黄色的中国花岗岩石板覆盖。

The flowerbeds are cut by this paving. The shown pictures were taken two weeks after planting. Many local rocks are used for walls and stone settings.

Main trees are sorbus intermedia, pinus nigra and acer neglectum annae. Shrubs like sambucus nigra and many mountain roses are used.

这条小路把花坛分开。展示的图片是植物种植两周后拍摄的。围墙和石头景观采用当地的岩石。主要种植的树木是花楸，欧洲黑松和欧亚槭。像欧洲接骨木这样的灌木和许多山地蔷也被使用。

The main idea of our concept is: the best garden is the nature around—let it be the best — do not try to work against it but enhance it by little impact.

我们的设计理念是：最好的花园就是四周的大自然，让它成为最好的，不要试图做不利于它的事儿，几乎不施加影响来提升花园的品质。

Redesign of Imchen Square and Waterfront Promenade

Imchen 广场和海滨长廊的重新设计

LOCATION：Berlin，Germany
项目地点：德国 柏林

AREA：10,000 m^2
面积：10 000 平方米

COMPLETION DATE：2007
完成时间：2007 年

PHOTOGRAPHER：Udo Dagenbach
摄影师：Udo Dagenbach

DESIGNER：Glasser and Dagenbach，Sabrina Schroeder，Silvia，Udo Dagenbach
设计师：Glasser and Dagenbach，Sabrina Schroeder，Silvia，Udo Dagenbach

DESIGN COMPANY：Glasser & Dagenbach Landscape Architects
设计公司：Glasser & Dagenbach Landscape Architects

Redesign of Imchen Square and Waterfront Promenade

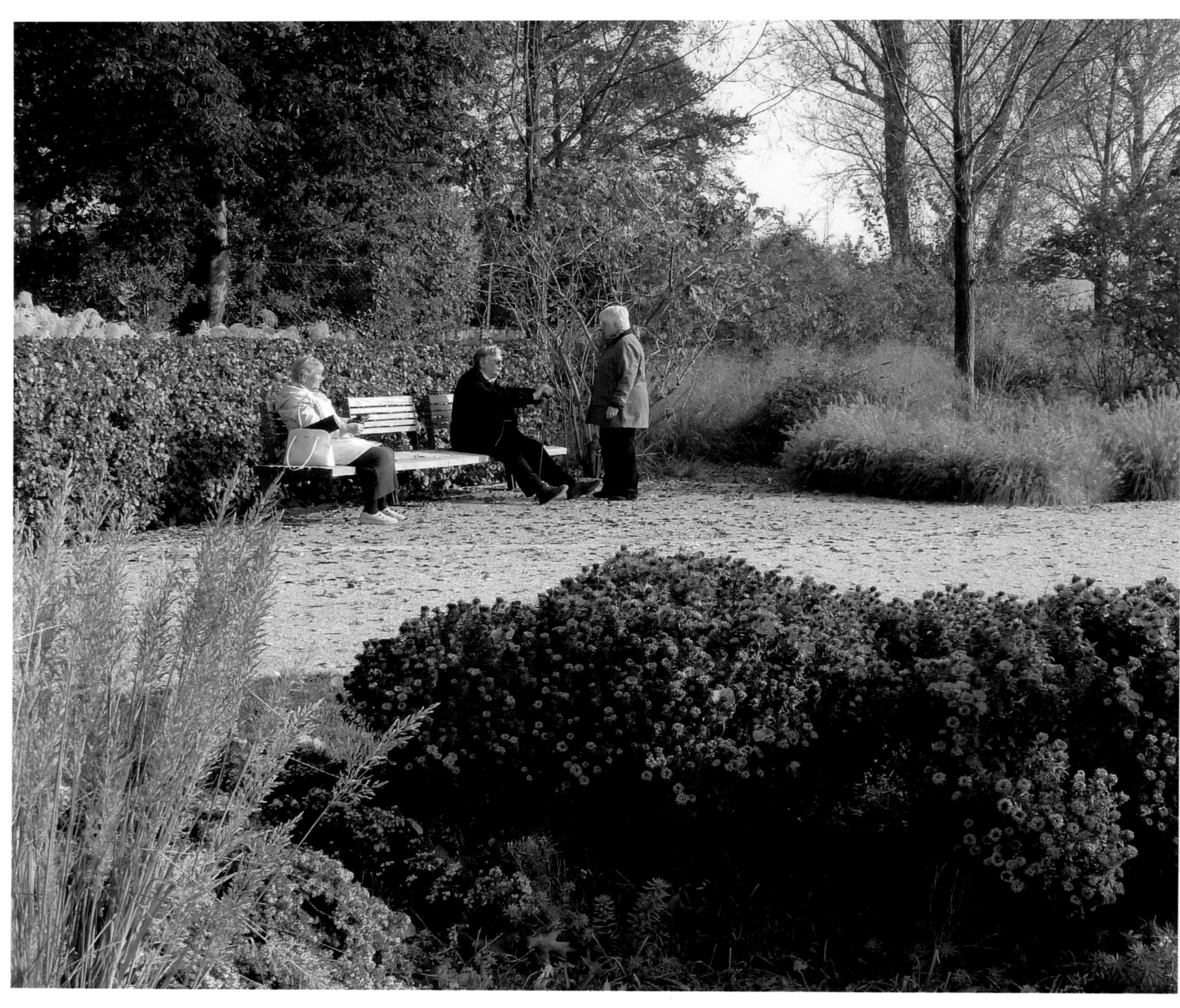

Imchen square is a modern square and shipping pier at the river Havel in Berlin. At the natural and soft coastal western side of the river Havel in Berlin-Spandau a very traditional square and shipping pier is located in the district Kladow—the so-called Imchen. Leftover structures from the early sixties and an overaged Tilia alley gave a poor impression of this very important local recreation area for Berlin's inhabitants.

Imchen 广场是柏林哈沃尔河边现代化的广场及船运码头。
在柏林 Spandau 哈沃河天然优美的西岸，一个非常传统的广场及码头位于 Kladow 区——就是所谓的 Imchen 广场。对于柏林的居民来说，20 世纪 60 年代早期留下的建筑和一个古老的椴树小径让人对这个当地重要的娱乐中心留下糟糕的印象。

The city council of Berlin-Spandau gave us the contract to do a strong redesign of coastline alley and square.

柏林 Spandau 委员会给我们这个合同，要求我们全面重新设计岸边的广场和小径。

Further the square should be prepared for a stronger use in future as Christmas market and place for summer party events.

进一步说，这个广场作为圣诞市场和夏天举办娱乐活动的地方应该准备好在未来发挥更大的作用。

Also the old boat loading ramp should be renovated and integrated in the new design.

而且，那个古老的装卸坡道应该翻新和修复，融入新的设计当中。

We decided to create a very reduced design which integrates both the fixed budget of 800,000 EUR and the multitasking requirements of many users.

我们决定创造一个简约的设计，用800 000欧元的固定预算满足许多使用者的多重要求。

Granite steps and platform follow the coastline. A canvas–shaped big lawn area flows to the river.
Smaller canvas–shaped areas with blood beach hedges and benches were adjusted to the walkway along the lawn to provide a good view to the river.

花岗岩石阶和平台分布在河岸边，如画布般的大草坪延伸到河边。
用海岸边的树篱地区围成了较小面积的风帆形状，然后在草坪两边放置长凳，从而绘出一幅美丽的滨河风景图。

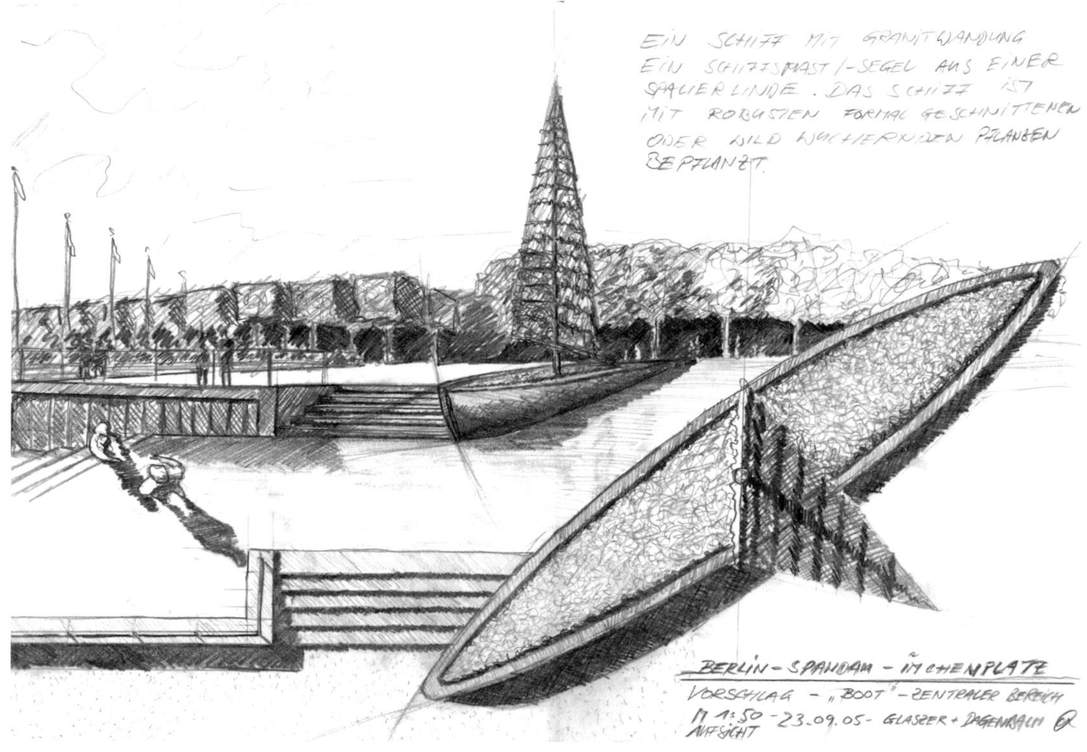

An over 40 m long curved plantbed for a prairie style perennial plantation is situated on one side of the lawn. Plantations of grass (like miscanthus and pennisetum) frame the red–leafed hedges.

草坪的一侧是一个超过40米长的弯曲的花坛，里面生长着草原地带多年生草本植物。草类（如芒草和狼尾草）成为红色树篱的边框。

To define the big space around the boat loading ramp we placed a boat made of blank shaped basalt stone blocks. The boat was planted with lavender.

为了界定船装卸坡道周围开阔的空间，我们放置了一个用毛坯玄武岩石块做成的小船，小船里种植了薰衣草。

The alley along the river was replanted with tilia cordata.

沿着河边的小路移植了欧洲小叶椴。

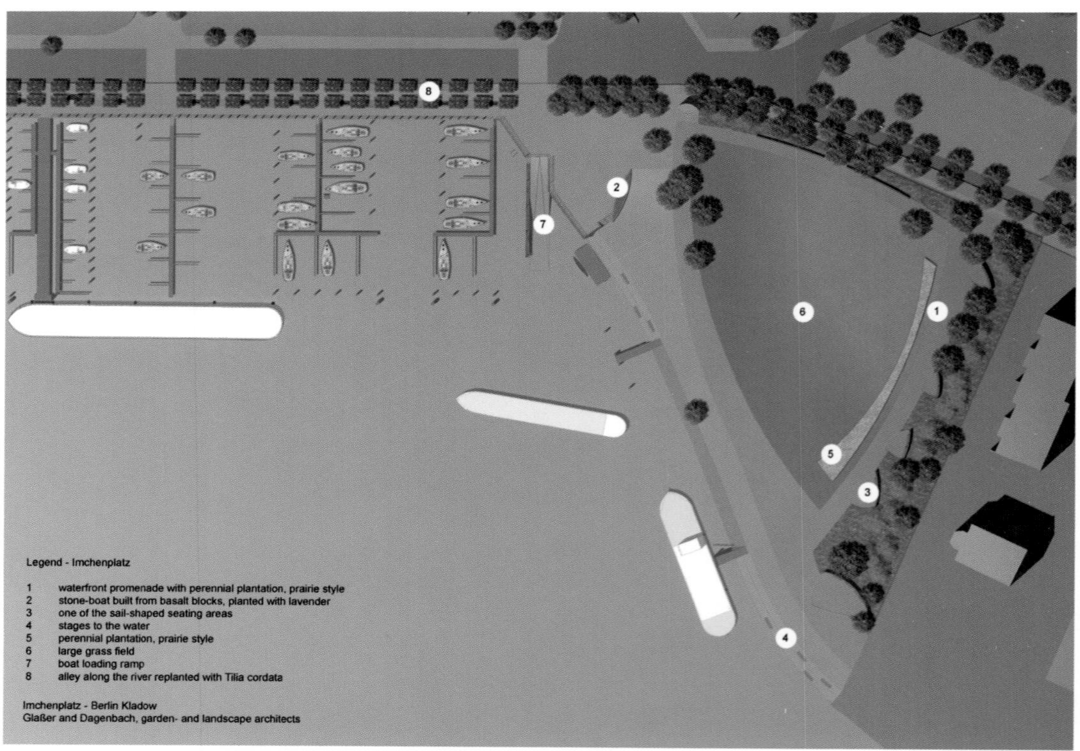

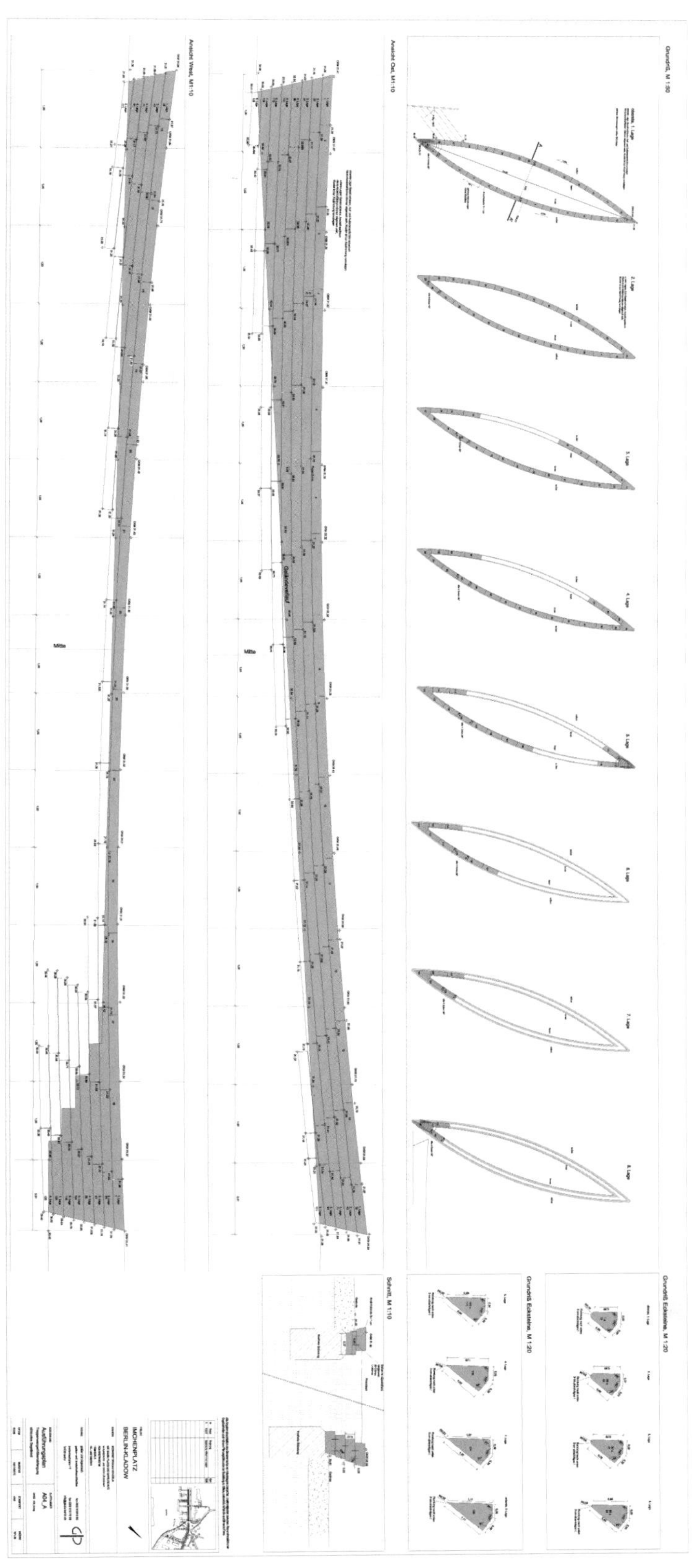

A Minimalistic Garden in a Forest in Vilnius Lithuania

立陶宛 维尔纽斯森林中的简约主义花园

LOCATION：Vilnius，Lithuania
项目地点：立陶宛 维尔纽斯

AREA：4,000 m^2
面积：4 000 平方米

COMPLETION DATE：2009
完成时间：2009 年

DESIGNER/LANDSCAPE ARCHITECT：Udo Dagenbach，Glasser and Dagenbach
设计师／景观设计师：Udo Dagenbach，Glasser and Dagenbach

ARCHITECT：Alfredas Trimonis
建筑师：Alfredas Trimonis

PHOTOGRAPHER：Udo Dagenbach
摄影师：Udo Dagenbach

A Minimalistic Garden in a Forest in Vilnius Lithuania

DESIGN COMPANY：
Glasser & Dagenbach
Landscape Architects
设计公司：
Glasser & Dagenbach
Landscape Architects

In 2008 we were asked to design the garden of a modernist style villa in a pine tree forest close to Vilnius, the capital of Lithuania. The building and the raw structures of the surrounding were designed by architect Alfredas Trimonis from HKT Architects in Hamburg.
The owner liked our sculptural projects and asked for a minimalistic approach to the garden design.

2008 年，我们应邀设计具有现代别墅风格的花园，项目地址在立陶宛首都维尔纽斯附近的一片松林中，建筑和周围的自然结构是由来自汉堡 HKT 建筑事务所的建筑师阿弗莱德·特利莫尼斯设计的。
该花园的主人喜欢我们的雕刻工程，要求简约的设计风格。

The whole spot takes its specific atmosphere from two elements: horizontal perfectly-maintained lawn and very high vertical pine tree stems. This creates a kind of melancholic, meditative mood which is very close to Japanese garden themes. The villa "sucks" in this forest mood using windows from the bottom to the ceiling. So you always see a part of the garden or forest from inside. In a wooden paving of an outdoor terrace the designers cut a rectangular hole. In this hole they arranged a cuboid sculpture representing the most possible reduction of a garden: half Jurassic marble and half clipped yew. In a buxus cuboid in the lawn they placed a spherical bronze calotte with various circular openings spreading like a star sky. At night it is lit from inside. At the backside of the house the designers arranged a spherical garden by creating a ball-shaped sculpture in the lawn between the vertical pine tree stems. One third consists of Jurassic marble and the other two thirds is again clipped yew.

整个设计从两方面呈现了特殊的氛围，水平的经过良好保护的草坪和笔直的松树树干，创造了一种忧郁的、让人沉思的氛围。这种氛围非常接近日本花园的主题。这个别墅利用从地面到天花板的窗户吸收了大森林的精气。因此，从屋内你总是能看到花园或森林的一部分。在一处户外平台的一条木质小路上，设计师裁出一个矩形的洞。在这个洞中，他们安排了一个立方体形的雕塑，展示了花园的简约：一半是侏罗纪大理石，一半是修整过的紫杉。在草坪上立方形的黄杨木中，他们放置了一个球形铜质圆顶，圆顶上分布有各种开孔，像星空一样。夜晚时，可从里面将其点亮。房屋的后面，在笔直的树干之间，设计者在草坪上创造了一个球形的雕塑从而形成了一个球形花园，三分之一是由侏罗纪大理石组成，其他三分之二是由修整过的紫杉组成。

Other parts of the garden close to the living room and sauna were designed as Japanese dry landscape garden with gravel, diabasic stones, 90 years old taxus cuspidata bonsais from Japan and amorphous clipped buxus sempervirens.

In the line of the Japanese garden another sculpture shaped as a discus was arranged. This time we wanted to have a garden element which loses any connection with gravity. That is why we created the top of the 1.4 m diameter discus of a levitating Jurassic marble stone. Below Taxus media was planted thus the complete shape of the discus is visible. The discus gets stainless steel stanchions which are shaped like stork feet anchored in concrete outside the discus axis. Stone and yew are connected symbolicly by a bronze disc.

花园靠近客厅和桑拿房的部分设计成日式旱风景花园，带有碎石和辉绿质的石头，从日本运来的有90年树龄的东北红豆杉，还有各式造型、修剪整齐的锦熟黄杨。在修建日式花园时，安放了一个铁饼形状的雕塑。这次我们想要制造一个失去重力的元素，那就是我们为什么要创造一个直径1.4米的铁饼，由漂浮在空中的侏罗纪大理石构成。在底部，种植了曼地亚红豆杉，因此可以看见铁饼完整的形状。铁饼有不锈钢支柱，外形像鹤脚，固定在铁饼中轴外侧的混凝土里。石头和紫杉用一个铜盘象征性地连接起来。

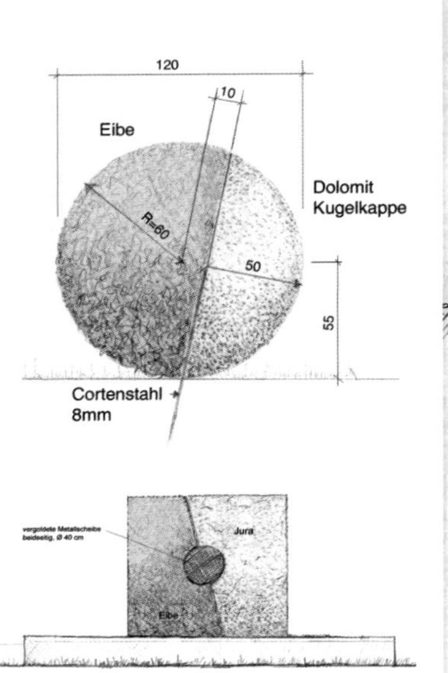

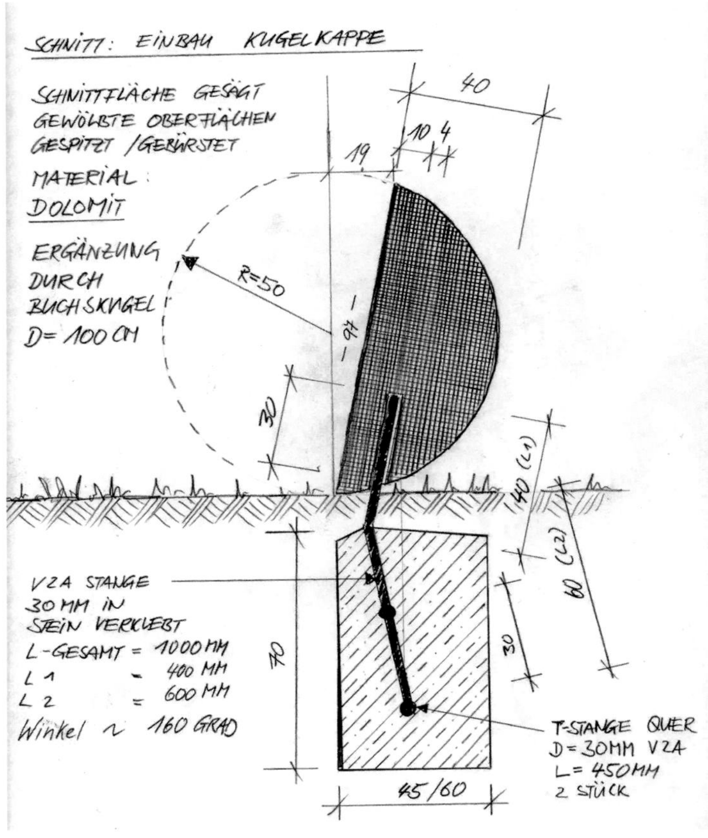

Charles Anderson / Atelier ps
landscape architecture urbanism
85 Columbia Street, Suite 101
Seattle, WA 98104, USA
Tel: 206. 322.0672
www.charlesanderson-atelierps.com

CHARLES ANDERSON

Principal of Charles Anderson | Atelier ps
Fellow, American Society of Landscape Architects (FASLA)
Adjunct Professor, Arizona State University, 2009–Present
MLA Harvard University, Cambridge, MA, USA 1985
Registered Landscape Architect in USA:
Alaska, Arizona, California, Montana, Nevada, Oregon, Washington
Charles Anderson is a licensed Landscape Architect with over 20 years of experience in projects ranging from neighborhood parks to New York's American Museum of Natural History. He has a strong background in public process and has completed many community projects. Anderson has a specific interest in expressionistic landscape restoration and in the development of urban ecologies. Museum and cultural institution work is a key interest of Charles, including projects such as the visitor centres at Mount St. Helens, the Seattle Art Museum's Olympic Sculpture Park, and Anchorage Museum of History and Art Expansion Project.

CHARLES ANDERSON | ATELIER PS

FIRM PROFILE

Charles Anderson | Atelier ps is a collaborative studio of passionate designers committed to creating artful, unique, site-specific, and well-crafted spaces.
We provide a shared vision of sustainable land stewardship, a high regard for function, and a dedication to quality. Our work is inherently multi-disciplinary – exploring landform changes and patterns within the context of natural science and engineering principles. We believe that environmental ethics and attention to detail go hand-in-hand with responsible selection of materials, minimization of waste and protection of our natural heritage.
The firm's principals have over twenty years of experience overseeing an extensive record of large and complex projects completed with internationally recognized architecture firms and artists. Landmark civic projects include the Seattle Art Museum's Olympic Sculpture Park on Seattle's downtown waterfront, the Arthur Ross Terrace at the Museum of Natural History in Central Park, New York City, Mount St. Helens National Volcanic Monument, a two-acre urban plaza and park for the Anchorage Museum of History and Art Expansion, the Lake Washington Environmental Education Sequence, and the International Peace Gardens at the border of Manitoba and North Dakota. Additionally, the firm's work includes over 25 community-based park projects that have involved extensive public participation, collaborations with community and neighborhood groups and public presentations.
In the course of creating solutions we draw on observations of ecological and social phenomena, processes, and latent relationships. Our sensitivity and dedication to thoroughly understanding each site allow us to create designs that balance natural system's requirements, aesthetic integrity and infrastructural constraints. We seek to maximize the potential of every project – while integrating art, ecological function and humanistic values. Our knowledge of native plants and natural systems is field-tested and thoughtfully employed to ensure that each design is appropriate to its place in the neighborhood, the city, and the region.
This commitment to synthesizing creative expression with functionality is complemented by an extensive background in all levels of the design and construction process: from site analysis and master planning to design development, construction documents, bidding and on-site construction observation. We provide innovative, efficient, and highly personalized professional services in the art and practice of landscape architecture – a comprehensive approach resulting in projects that stand out over time.

Founded 1990
Seattle, Washington

Arthur Ross Terrace at the American Museum of Natural History

美国自然历史博物馆的亚瑟·罗斯平台

LOCATION: New York, USA
项目地点：美国 纽约

AREA: 8,094 m^2
面积：8 094 平方米

DESIGN COMPANY: Charles Anderson | Atelier ps
设计公司：查尔斯安德森景观建筑公司

Arthur Ross Terrace at the American Museum of Natural History

Since construction of the Planetarium and new parking garage had already begun, one of the biggest challenges was to create a design to fit the tree pits and infrastructure of the previous design for the terrace. In retrospect, the process at times felt like attempting to fit a likeness of Julia Robert's face over Walter Mathieu's—a task not easily done.

由于天文台和新停车场的开建，我们所面临的最大挑战之一便是要创建一个能与上次平台设计中安置的树坑和基础设施相适宜的新设计。如果回顾一下我们这一阶段的进展，感觉就像是在沃尔特·马修的脸上重塑一张像茱莉亚·罗伯茨的脸。这并不是一项轻松容易的任务。

The rooftop terrace was designed in tandem with the construction of the Rose and Priest Centre for Earth and Space, which replaced the aging Hayden Planetarium, a parking lot and museum service area. One part of the new construction was the Arthur Ross Terrace, which created an acre of new semi-public open space. The terrace, a multi-purpose urban plaza constructed over a new parking garage welcomes the public and visitors to the museum, hosts special events, and provides outdoor educational spaces for museum patrons, school children and the general public. The design of the terrace links the historic, traditional appearance and function of the American Museum of Natural History with the modern design and gleaming materials of the Rose Centre for Earth and Space. The physical representation of this concept was inspired by an illustration of the multiple, conical shadows cast by a moon during an eclipse. The futuristic, floating sphere of the Planetarium becomes an eclipsing moon that cast "moon shadows" carrying the sphere's celestial presence across the terrace.

与屋顶平台设计同步进行的还有玫瑰与牧师中心的建设，此中心用于地球和宇宙展区的展览，代替了陈旧的用做停车场和博物馆服务区的海登天文台。新建项目中的一部分便是亚瑟·罗斯天台，天台为博物馆提供了一英亩见方的半公共露天空地。天台成为位于新建停车场上方的多功能城市广场，喜迎公众和博物馆参观者们的到来，此外，广场还可用于举办特别活动，为博物馆的赞助商、学龄儿童和广大市民提供户外教育场地。天台的设计融合了美国自然历史博物馆颇具历史意义的、传统的外观和功能，在地球和宇宙展区展览的玫瑰中心采用了现代设计和发光材料。表达这一概念的灵感来源于月食时月亮投射下的多重圆柱状阴影。天文台设计中具有未来主义风格的、浮动的球体就像月食中的月亮一样，在天台上投射下拥有美丽仪态的“月亮影子”。

The terrace serves as a perfect setting for viewing the exquisite new Planetarium building, with a glass cube housing an iconic sphere, which redefined the image of the 128–year old institution while respecting its landmark designation. The terrace also redefines the landscape vocabulary for the traditional grounds by introducing new materials in a setting of metaphors and simplicity. The terrace is a space where prehistoric Ginkgo trees meet the stars, creating a place of reflection, learning and rest. Visitors are prompted to contemplate our connection to the earth, as well as explore the wonders of space.

平台还是观赏精美绝伦的新天文台的绝佳场所。天文台使用标志性的星体外罩——晶莹的玻璃立方体。这一设计不仅重新定义了拥有 128 年历史的博物馆的形象，同时还出色地完成了其成为地标建筑的使命。同时，平台的设计还颠覆了传统意义上“风景”这个词汇的概念，使用了一系列具有隐喻性和简洁性的新材料。另外，在平台上，古老的史前银杏树与星际相接，创造了一片可供人们沉思、学习和休憩的场所。在这里，参观者们将会兴致勃勃地研究人类自身与地球的联系以及探索太空的奥秘。

ARTHUR ROSS TERRACE

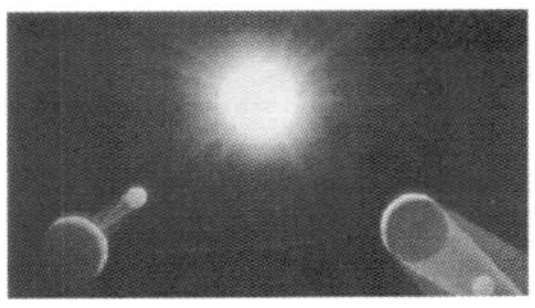

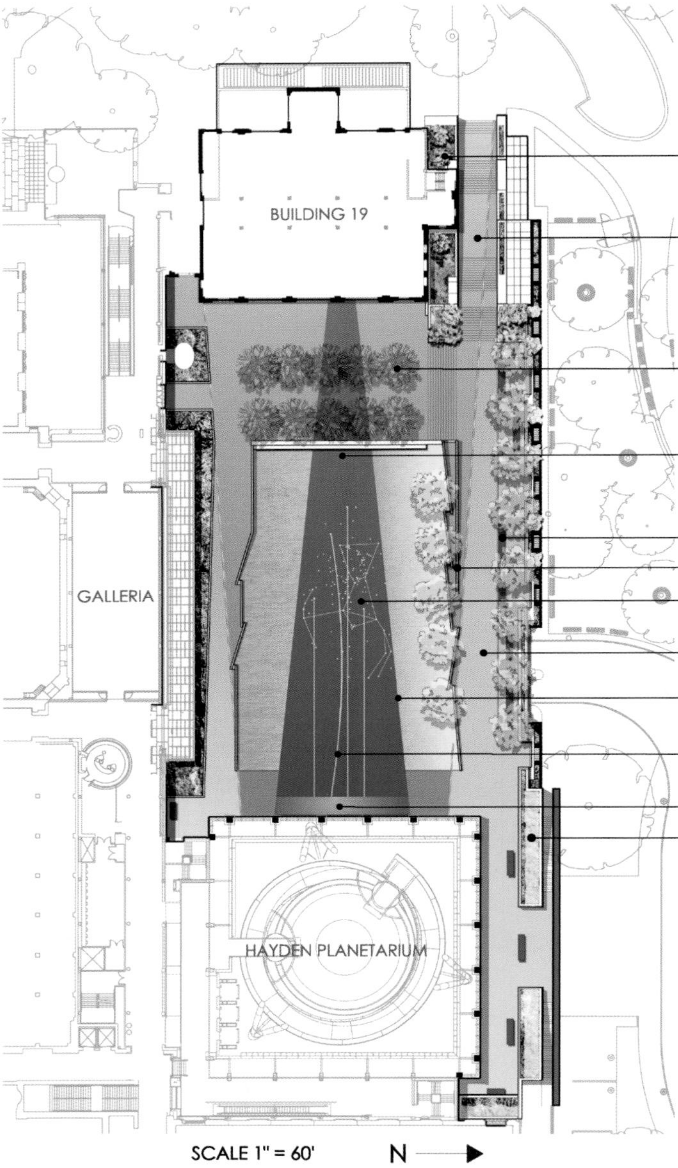

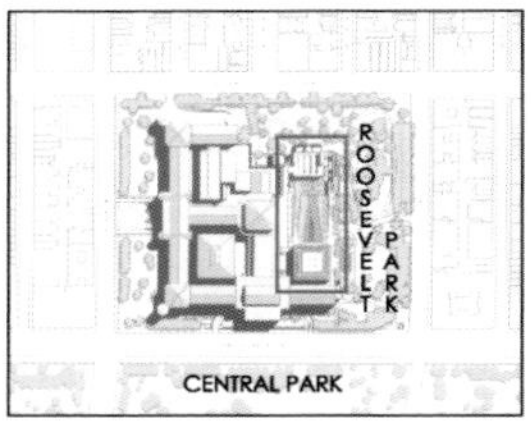

BUILDING 19

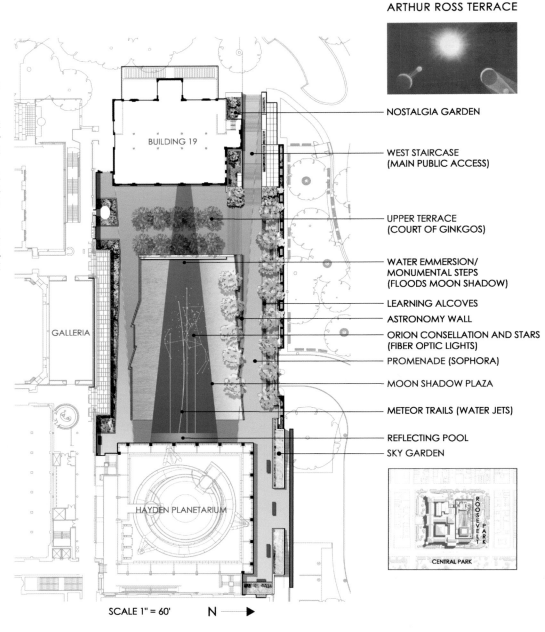

GALLERIA

HAYDEN PLANETARIUM

NOSTALGIA GARDEN

WEST STAIRCASE (MAIN PUBLIC ACCESS)

UPPER TERRACE (COURT OF GINKGOS)

WATER EMMERSION/ MONUMENTAL STEPS (FLOODS MOON SHADOW)

LEARNING ALCOVES

ASTRONOMY WALL

ORION CONSELLATION AND STARS (FIBER OPTIC LIGHTS)

PROMENADE (SOPHORA)

MOON SHADOW PLAZA

METEOR TRAILS (WATER JETS)

REFLECTING POOL

SKY GARDEN

ROOSEVELT PARK

CENTRAL PARK

SCALE 1" = 60'

N

FREDERICK PHINEAS & SANDRA PRIEST ROSE

Bellevue Collection

贝尔维尤收藏馆

LOCATION：Bellevae，usa
项目地点：美国 贝尔维尤

AREA：10,927 m^2
面积：10 927 平方米

LEAD DESIGNER：Larry M. Smart，ASLA
设计总监：Larry M. Smart，ASLA

PHOTOGRAPHER：Larry M. Smart
摄影师：Larry M. Smart

DESIGN PERIOD：Ongoing since 2005
设计时间：2005 年至今

COLLABORATION：Sclater Partners Architects
合作者：Sclater Partners Architects

DESIGN COMPANY：Charles Anderson | Atelier ps
设计公司：查尔斯安德森景观建筑公司

Bellevue Collection

With a savvy owner and development team with a clear vision to help craft Bellevue's future for a livable vibrant downtown, we have shared a long working relationship with Kemper Development Company based on trust and the shared commitment to do things right. From urban streetscape and garden court design, to site adaptation for sculpture and art installations and water feature design, we are proud to have collaborated on over 20 projects with Kemper Development, contributing to realizing the owners' vision for an active, livable urban centre.

拥有一位精明的领头人和一支强劲的开发队伍，外加一个明确的设想，我们的团队希望能将贝尔维尤的将来规划成为一个宜居且充满活力的城市中心区。基于彼此的信任和把事情做对的承诺，我们和肯铂发展公司进行了长期的合作。从城市街景和花园庭院的设计，到雕塑作品和艺术设施的现场安置以及水景的设计，我们很荣幸与肯铂发展公司拥有超过 20 个项目的合作，为实现业主创造一个积极、宜居的城市中心的设想做出自己的贡献。

EAST COURT

A richly-planted courtyard, framed on two sides by mature European Hornbeam, this garden is a verdant place of respite for Bellevue Place tenants, clients, and visitors as well as an elegant venue for more formal events. The lawn, engineered over a parking structure, is traversed by concrete planks that terminate in two gathering points. Perennial beds, flanked by long LPE benches edged with core-ten steel, overflow with soft-textured perennial plantings, ferns and ornamental grasses.

东部庭院

东部庭院植被丰富，庭院的两边种植着成熟的欧洲角树。这所葱郁的花园可为贝尔维尤广场的租户、客户以及参观者提供一个休憩的场所，同时还是一个可举办更多正式活动的优雅场地。置于停车场结构上方的草坪，生机勃勃、长势良好，上面遍布着弧形的实木木板，由同一点射线般地散射出，最终又归集于另一点。种有多年生植物的花坛，满溢着长有柔软纤维的多年生植物、蕨类植物和观赏类花草。在花坛的两侧安放着带有钢边缘的聚乙烯长椅。

STREETSCAPE

Prior to the adoption of the City's "Green Streets" program, design and development for the Bellevue Collection retail frontages focused on realizing welcome and comfortable outdoor pedestrian zones. Featuring broad sidewalks separated from adjacent surface streets by generous street edge plantings, pedestrian improvements typically include staggered or offset granite blocks, concrete seating pockets and pedestrian scaled lighting. To date, the Bellevue Collection streetscape development comprises approximately eight linear blocks of improvements, including the Westin Hotels' entry frontage and completion of the City's N.E. 6th Pedestrian Corridor.

街景

在采用城市“绿色街道”项目之前，对贝

尔维尤中零售商店前街道的设计专注于打造宾至如归、舒适的户外行人专用区。极具特点的宽阔的人行道，通过在路边种植的大量植物与邻近的外侧街道分割开来。人行道的改善主要包括在街道上零散地或是平衡地放置花岗岩石块，安置混凝土的座椅和鳞状行人照明装置。迄今为止，贝尔维尤收藏街景的开发包括大约 8 个线性排列的街区的改进，其中包括威斯丁酒店入口临街的设计和城市东北第六人行走廊的竣工。

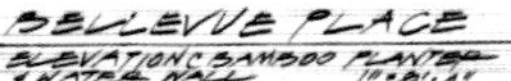

WINTER GARDEN

Expansion of the Bellevue Hyatt Regency Hotel includes renovation of the Bellevue Place Winter Garden, a gathering place and hub to the complex of hotel, restaurants, and office tower. A basalt clad and blackened steel water feature diffuses sound within the large atrium space, providing a tranquil edge to the elevated registration area of the hotel and focal point for adjacent quiet seating, meeting and circulation space. Groves of timber bamboo create a comfortable canopy to floor level activities while providing a vertical accent beneath the glazed dome ceiling.

冬景花园

贝尔维尤凯悦大酒店的扩建包括对贝尔维尤广场冬景花园的整修。冬景花园是酒店、餐厅和办公楼混合体的聚会场所和中心。玄武岩和黑钢片覆盖的水幕景观在广大的中庭内弥漫，为平台上宾馆登记区的旅客们提供了一个僻静的角落，同时，水幕景观还成为毗邻地带安静的休闲、会见和通道观赏的焦点。一片竹林形成了一张舒适的华盖，人们在竹影的遮盖下举办各式各样的活动；同时，在玻璃制成的穹顶下，在建筑美学上，竹林的存在还提供了竖向的线条。

FOUNTAIN COURT

Anchoring Bellevue Squares' primary east pedestrian entry and terminus to the City's new N.E. 6th Pedestrian Corridor, the fountain within the space was overlooked and dated. In concert with street frontage landscape improvements we were charged with updating the existing feature utilizing the existing primary mechanical systems and basic footprint until a major redevelopment of the Squares' entry and court space was directed. The resulting redesign successfully updated the character of the water effect, level of finish and materials as well as the courts' function as gathering and meeting space.

喷泉庭院

贝尔维尤广场东人行道入口已进行初步建设，到达城市新东北第六人行走廊的总站已经破土动工，但是此空间内喷泉景观的设计与建设却被忽略以及延期。除了负责改善街道邻街景观，我们还负责更新那些基于现存初级机械系统和基本框架所实现的功能，直到接到重新开发广场的入口和庭院空间的指示。这次的重新设计成功地更新了水景的效果、提高了景观表面的光洁度和所用材料档次，同时，还升级了庭院作为聚集和会面场地的功能。

Denver Arts and Crafts House

丹佛市艺术馆和工艺品楼

LOCATION：Colorado，USA
项目地点：美国 科罗拉多

AREA：14,164 m^2
景观面积：14 164 平方米

PHOTOGRAPHER：Charles Anderson/Atelier ps
摄影师：Charles Anderson/Atelier ps

Landscape Architect：Charles Anderson，FASLA
景观设计师：Charles Anderson，FASLA

COLLABORATION：Jim Olson，Olson Kundig Architects
合作者： Jim Olson，Olson Kundig Architects

Denver Arts and Crafts House

Denver is a difficult place for trees to grow—its prairie character and the extremes of the weather make it suitable for only the hardiest of plants. Working in this environment called for us to protect a number of existing large trees and import a whole new forest of locally appropriate aspen trees.

丹佛是一个非常不适宜树木生长的地方。它的北美大草原性气候和恶劣的天气状况使得此处只适合那些最顽强植物的生长。在这种环境下工作，需要我们保护若干现存的大树和进口一整片适应当地状况的白桦树树林。

The aspen grove embraces the rear and side yard of the site, while at the front open grasses and meadows make every view from inside of the home a distinct experience. Native and native-compatible species comingle here, and as the landscape nears the home, a more exotic manicured landscape prevails, complementing the extensive "civilized wilds."

白桦树丛环绕着场地的后院和侧院，场地前方的开阔地带覆盖着草坪和草地，无论从任何角度，场地都为居家的观赏者提供了独一无二的景致。本地的和适宜本地生长的物种在此共同生长。由于此景观靠近居民区，因此修建整理一个更具异域风情的景致会更受欢迎，此外，景观还为广阔的“文明荒野”增添了一份景致。

We retained the remnants of a gigantic waterfall feature in the new forest, partly to reveal the site's history, but mainly because a fox lived in its rocks and, within them, grew two very old Pinion Pine trees. The client loves rabbits and has an extensive collection of hare art and sculpture. An enormous sculpture of a hare by Barry Flanagan stands in the window of the grand foyer enticing the resident fox in the garden.

在新种的白桦树林中，我们保留了巨大的瀑布景观的残留部分，这样做一部分是由于要展示场地古老的历史，但最主要是因为在场地里的岩洞中居住着的一只狐狸和为另外两棵非常古老的齿轮松树提供水源。我们的客户非常喜欢兔子，他收集了大量的野兔工艺品和塑像。在宏伟的门厅前就矗立着一座巴里·弗拉纳根的作品——一只巨大的野兔雕像，时刻引诱着居住在花园里的狐狸前来猎食。

The design team brought their respective experiences and concerns to the table, but always worked to make this a landscape and home entirely for the client. Working closely with the client, we sited the new home, determined where guests would arrive and be entertained, and arranged a landscape with great variety and a sense of humor. Through growth, the landscape continues to be a design in progress. The most significant aspect of this project is the feeling that it is a special, place that seems much bigger than it is, and yet is intimate and comfortable.

设计队人员根据各自的经验提出了他们的担心，但是，他们一直努力为客户打造一个景观和住房一体的项目。通过与客户的密切沟通，选定了新住房的地址，确定了应当在什么地方迎接和款待客人，并且将景观设计得即富有变化又不失幽默。景观日益丰富，这显然是一个仍在进行中的设计。这个项目最意味深长的一点是它令人觉得很特别，看上去比实际面积大很多，而且令人感觉很亲切很舒适。

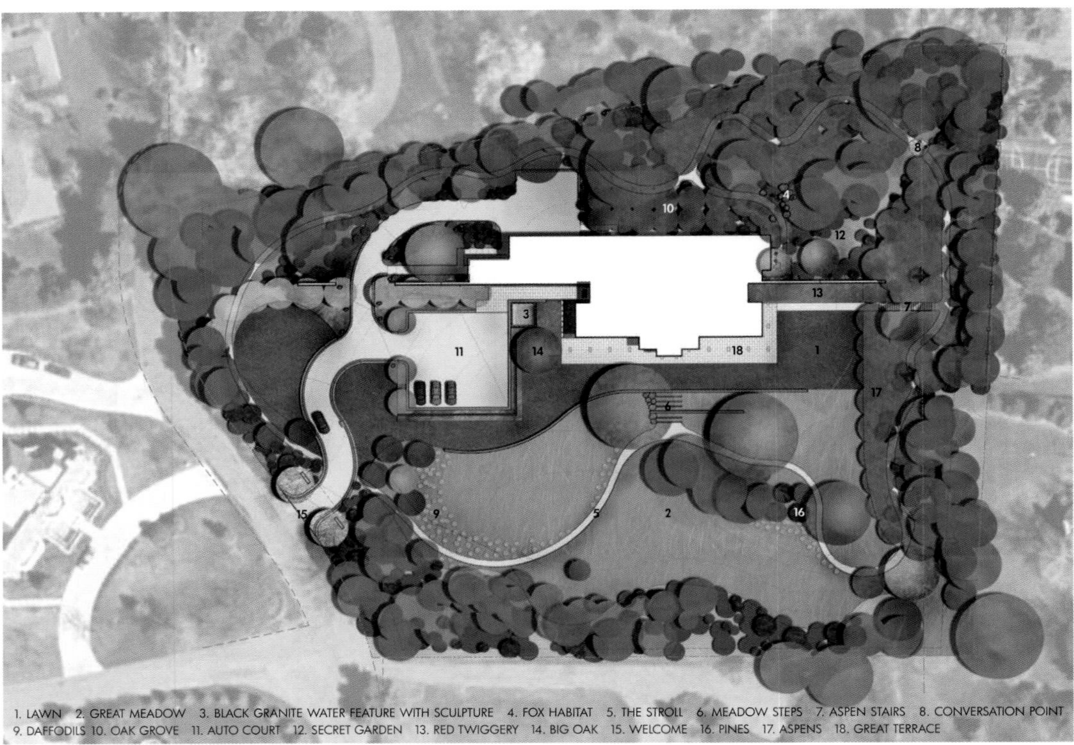

PA - 265 - 01 Master Site Plan

Hill House (Lil Big House)

山楼（李尔大楼）

LOCATION：Washington，USA
项目地点：美国 华盛顿

AREA：84 m²
景观面积：84 平方米

PHOTOGRAPHER：Philip Vogelzang
摄影师：Philip Vogelzang

DESIGNER：Charles Anderson，FASLA
设计师：Charles Anderson，FASLA

COLLABORATION：Tom Kundig，Olson Kundig Architects
合作者： Tom Kundig，Olson Kundig Architects

DESIGN COMPANY：Charles Anderson | Atelier ps
设计公司：查尔斯安德森景观建筑公司

Hill House (Lil Big House)

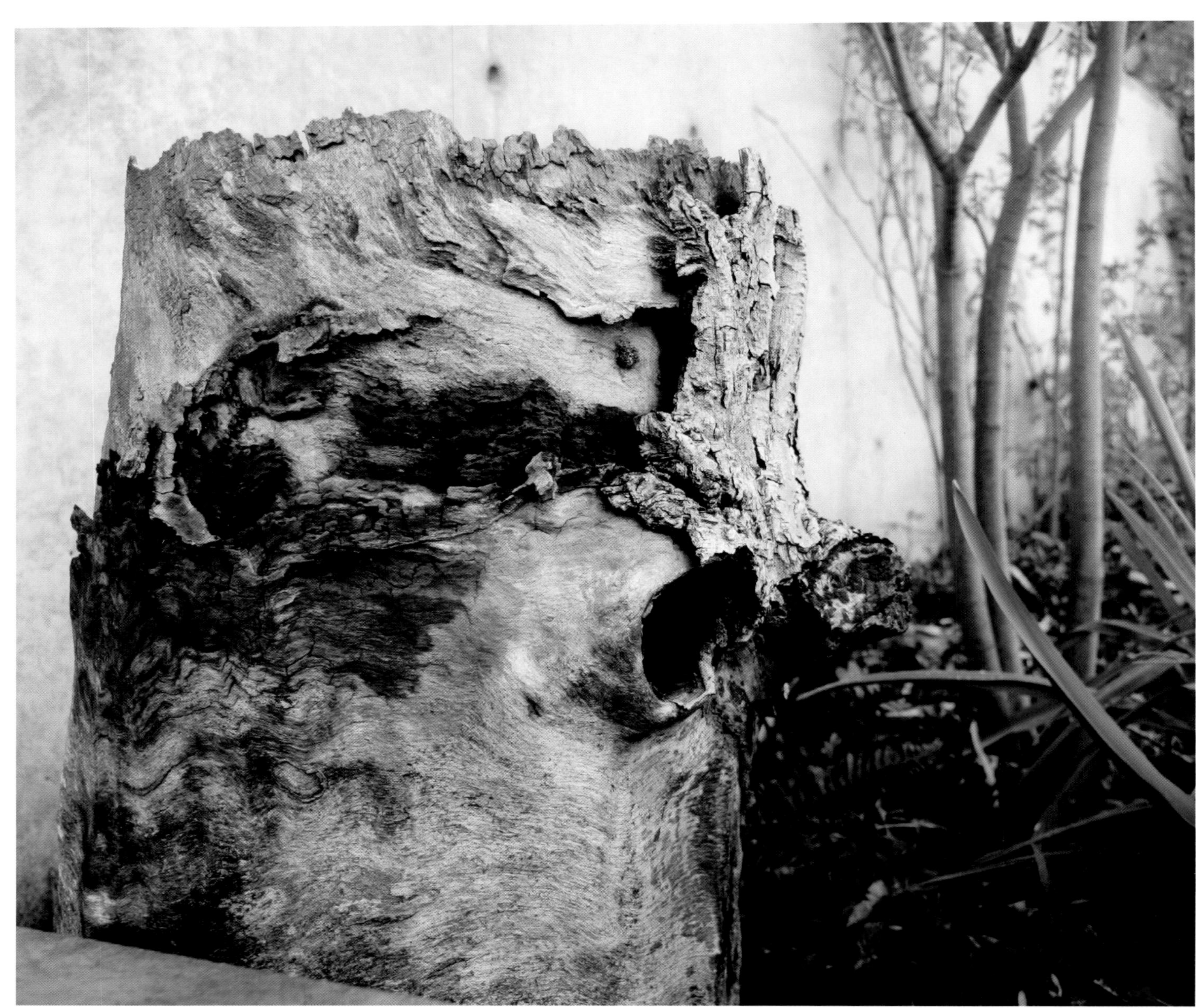

Perched atop Queen Anne Hill and overlooking Puget Sound, this contemporary metal-clad home is set amid a landscape of all native plants except for those in the rooftop vegetable garden. The residence and surrounding grounds reflect the values and aspirations of the owners to achieve a balance of urban building aesthetics and horticultural utility while capitalizing on an amazing view, which are realized with the respect, help and encouragement of the architects and the landscape architects.

坐落于安妮女王山的顶端，俯瞰着皮吉特海峡，这座被金属覆盖的现代化房屋置身于大片当地植物组成的风景之中。住宅和周围庭院的设计反映了业主们在利用令人惊叹的风景的同时，实现了城市建筑美学和园艺实效艺术平衡的价值观和愿望，并得到了建筑师和景观设计师们的尊重、帮助和鼓励。

Space constraints have been transformed into landscape opportunities at different levels of the site. Taking advantage of narrow and unusually steep city lot, the 242 m^2 house acts and appears to be double that size. Although the surrounding landscape is only 84 m^2, it expands both into the house and out to surrounding landscape to give the feeling of a small estate. With its saturated colors and raw sculptural materials palette of oxized steel and natural stone and concrete in complementary tones, the house looks and feels as if it were rising out of the earth. The gardens are planted in Big Leaf and Vine Maples, Madrone, Salal, Snowberry, Tall Oregon Grape, Bunchberry and Western Trillium, which are a few of the "natives" that share the smallest of lots and soften the building mass.

空间的束缚被转化成在场地建设不同层次景观的契机。设计充分利用了狭窄而且异常陡直的城市空地，使242平方米的房屋看上去有其面积的两倍大。虽然，周边景观的面积只有84平方米，但向周边景观和房屋内部的延伸拓展使它看起来就像一个小庄园。其充满饱和度的色彩和原始雕塑材料调色板般相互补充的氧化钢、天然石头和混凝土的颜色使房屋看上去像是从地面上升起来的。花园里种着大叶、藤槭、石南科小灌木、沙龙白珠树、美洲针金葛、高俄勒冈葡、御膳橘和西方延龄草，还有能在狭小的城市土地里共同生长并且能软化建筑土地的极少数本地植物。

The front entrance pulls a visitor through the planted terrace to the canyon-like staircase. At the top of the steps one turns and is rewarded with spectacular views of Puget Sound and mountain views. From the alley side one enters through the lowest floor of the house by way of a verdant wetland grotto filled with Giant Horsetail, Skunk Cabbage, and Slough Sedge. Above the grotto and spanning a water feature below is a metal grated walkway. Spaces under grated ramps and stairs are filled with large boulders, ferns, and mosses. Off the back of the house, at the entry level, there is a protected garden of refuge. The side yard is a native plant nursery that will host young plants to be used for restoration efforts elsewhere in the city. The roof of the garage acts as a perch for growing vegetables and herbs.

前入口将拜访者从一块植物覆盖着的平台引入一段峡谷般的楼梯。在楼梯的最顶端，如果回眸而望，则可欣赏到皮吉特海峡和远处群山壮丽的景色。如果从小巷这边进入，便需通过房屋最低层，经由一个葱郁的长满巨型马尾草、臭菘和斯劳莎草的湿地岩穴进入。在岩穴上方，横跨水景的是一个金属格栅的人行道。格栅的斜坡和楼梯下的空间充满了巨石、蕨类和苔藓植物。在房屋的后面，入口处，有一所用于保护物种的庇护花园。其侧院是一个培育本地植物的苗圃，向城市其他地方供应用于恢复绿化的种苗。车库的顶部用于种植蔬菜和药草。

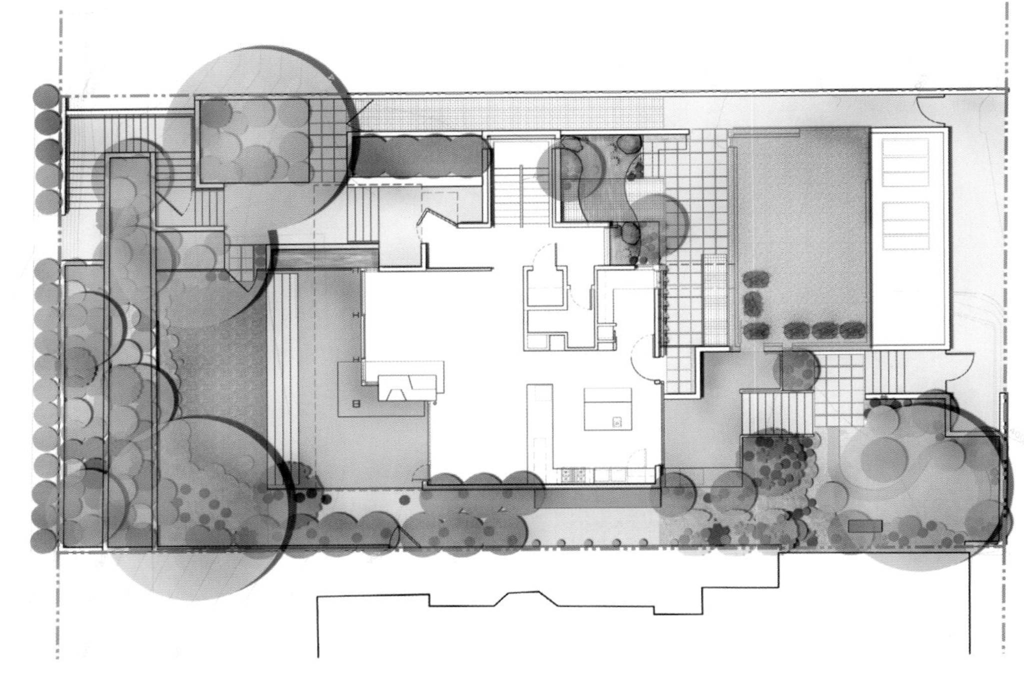

This project is a demonstrational landscape of native plants and wildlife habitat as well as a reflection of the owners' passion for ecology, art, and architecture. It is a garden of stored rainwater, a series of native plant communities learning how to be "urban" as well as someone's home. This is a contemporary landscape where the understanding of what is inside and what is outside is blurred and shared. From this property you will see an eagle fly overhead at the same time as you watch black capped chickadees flicker and rummage the grounds for food at your feet. You can see the lights of the city at the same time as you see the blackness of the water beyond. It's a place of contrasts balanced with harmony that truly reflects the owners' belief in a healthier way of life, not just human life, but one inclusive of other lives as well.

本项目是本地植物和野生生境的示范景观，同时也反映出业主对生态、艺术与建筑的热情。这是一个雨水蓄积公园，里面有很多本地植物群落，试图在适应建筑住宅的同时，打造出“城市”的特征。项目采用当代景观，模糊了室内和室外的界限。在这里，你可以看到老鹰在上空盘旋，也可以看到黑顶的山雀扑动着翅膀在地上寻找食物；你可以看到城市的灯光，也可以看到黑夜中远处的水面。这里是和谐与对比的统一，反映出了业主对健康生活方式的信念，不仅是人类的生活方式，而且也包括其他生物的生活方式。

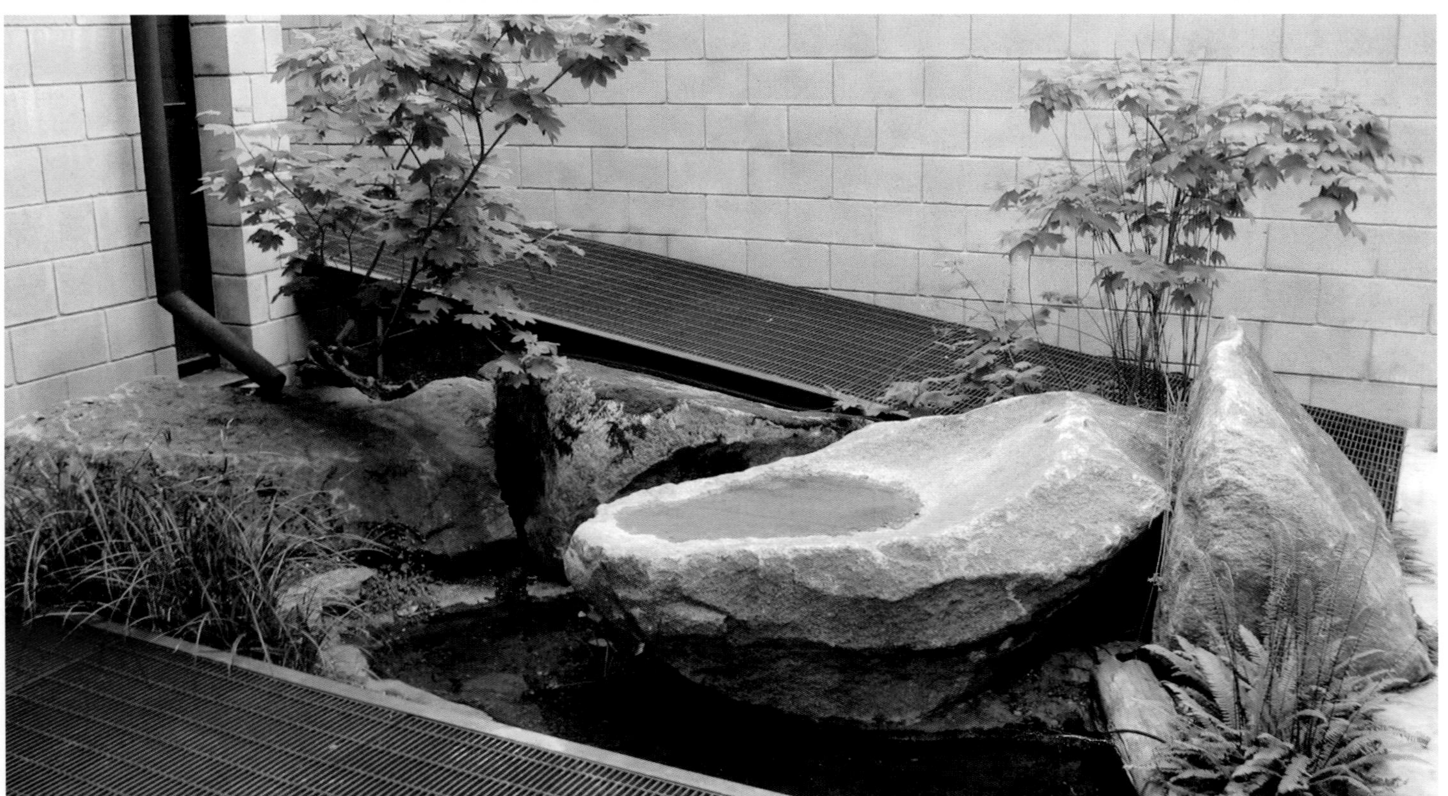

Island Square

岛屿广场

LOCATION: Seattle, USA
项目地点：美国 西雅图

AREA: 10,592 m^2
面积：10 592 平方米

COMPLETION DATE: 2008
完成时间：2008 年

PHOTOGRAPHER: Philip Vogelzang
摄影师：Philip Vogelzang

DESIGNER: Charles Anderson, FASLA
设计师：Charles Anderson, FASLA

COLLABORATION: Mithun, Jim Bodoia
合作者：Mithun, Jim Bodoia

DESIGN COMPANY: Charles Anderson | Atelier ps
设计公司：查尔斯安德森景观建筑公司

Island Square

The project features two large internal courtyards designed primarily for use by residents, both of which are engineered using post-tensioned concrete lid slab and a pedestal paver system over the parking garage below. The focal point of the Spa Courtyard is a two-tiered, heated waterfall pool and outdoor shower tiled in a brilliant, Caribbean blue that seems to glow through the foliage of the surrounding Bamboo. The upper spa pool is covered by a sizeable glue-lam beam and Kalwall arbor structure, which also doubles as a roof for the attractive pool pavilion below, which conceals and protects the mechanical equipment. The adjacent lounge area is surrounded by plantings of Bamboo as well as a rockery garden containing flowering Dogwood trees underplanted with Japanese Forest Grass and David's Viburnum. These elements combined with the sound of falling water create a soothing and rejuvenating atmosphere that defines the courtyard.

此项目以两个大型内部庭院为特色，庭院主要为满足居民使用而设计。这两个庭院都位于车库上方，使用后张法混凝土盖状厚板和一个基座摊铺系统。温泉庭院的设计焦点是一座两层的浴室，内有加热瀑布池和户外淋浴。浴室的墙壁贴着鲜艳的、加勒比蓝色的磁瓦，在周围摇曳的竹叶中闪闪发光。上层温泉池上覆盖着一个面积相当大的胶林梁和酷沃凉亭结构。这一结构同时也用做下面具有吸引力的水池凉亭的屋顶，隐藏和保护着一些机械设备。毗邻的休息区周围围绕着竹林和一座石景园。石景园里种植着开花的山茱萸、日本森林草和大卫琼花。这些设计元素，伴随着潺潺的流水声，能为游客们营造和提供一种舒缓的、令人恢复活力的氛围，这也正是我们建造庭院的初衷。

The second private courtyard—the Arbor Court is a spacious and luxurious outdoor living room. The wood arbor is both a focal point and the destination at the west end, while the raised, cast-in-place planters form a variety of seating, gathering and conversation spaces amongst the trees. The planters seem to overflow with plantings of ornamental grasses, sedges, succulents and vines, while the multi-stem Dogwood and Japanese Maple trees offer a sense of privacy and separation between seating areas and unit entries. The planters step down in elevation at the far end of the courtyard, creating a sense of welcome as one enters the space from the Sun Court.

第二所私人庭院——凉亭庭院，是一个宽敞的、豪华的户外起居室。木制的凉亭不仅是焦点景色，而且还是庭院西端的终点。而凸起的、就地浇铸的培植器皿在树林中组成了多种多样的休息区、聚会区和交谈区。这些培植器皿里充满着观赏性草、莎草、肉质植物和葡萄藤，其中种植的多干山茱萸和日本枫叶草则为庭院提供了一种私密感，并将休息区和入口区分割开来。这些培植器皿一个比一个更高地向庭院的尽头蜿蜒而去，仿佛在欢迎由太阳庭院进入的宾客们。

The Flower Court faces west, and captures views toward the beautifully-planted garden median at 78th Avenue. It incorporates an information kiosk and the parterre planting of Blue Oat Grass, New Zealand Flax and Lavender and accents the adjacent pedestrian arcade. It features colorful containers and granite seating blocks, as well as outdoor patio spaces for use by retail tenants which are shaded by Honey Locust Trees.

鲜花庭院坐东朝西，在此花园中能观赏到位于78号大街中段经过精美布置过的花园的景色。鲜花花园还合并了信息台和种有蓝色燕麦草、新西兰亚麻和薰衣草的花坛。这个花坛突出了毗邻的人行拱廊。鲜花庭院的特色是色彩斑斓的容器、花岗岩的座位块和皂荚树遮盖的供零售租户使用的户外庭院空间。

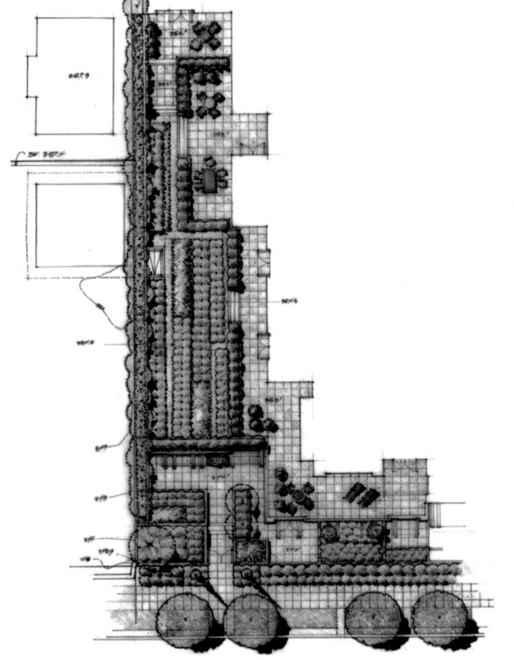

The Water Court also faces west and orients visitors with an informational kiosk. The wall supporting the staircase leading up to the Spa Courtyard doubles as a water wall for a raised pool below. The essential character of this courtyard plaza is a calm and shady oasis, with the sound of traffic masked by the sound of water falling.

水苑同样坐东朝西，并且内置一信息台为游客提供向导服务。水苑的院墙支撑着通向温泉庭院的楼梯，同时还是下面一个凸起水池的水墙。这座庭园广场最本质的特征就是它是一座平静且阴凉的避风港，嘈杂的交通声被潺潺的流水声所掩盖。

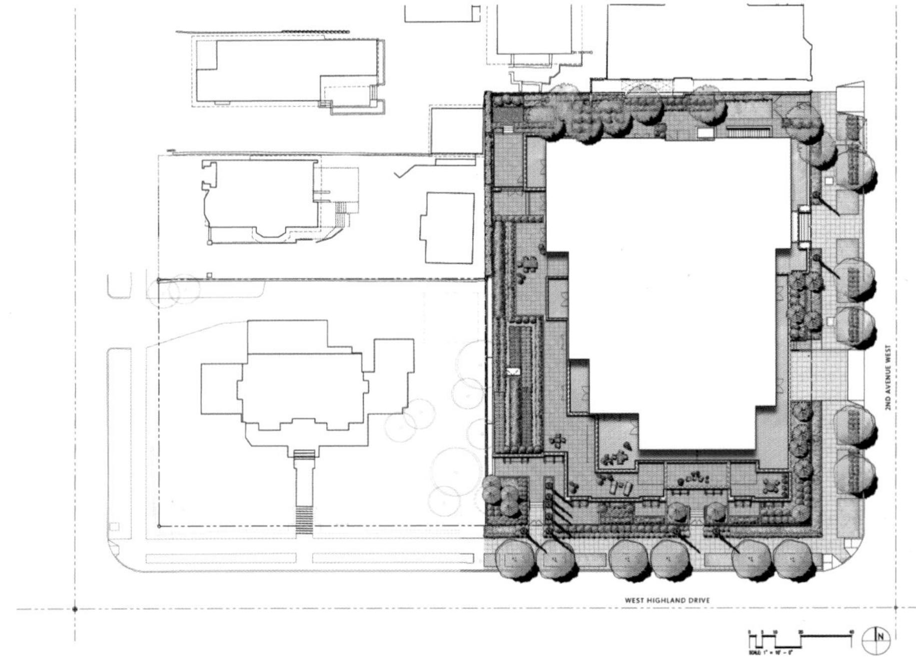

The Sculpture Court sits prominently at the southwest corner of the property, and is tied strongly to the context of the site, through the use of a similar vocabulary of artwork to the existing streetscape on the opposite corner. A cast-in-place sculpture by the local artist sits atop a rough-hewn granite slab, and is surrounded by a planting of Lavender and Japanese Silver Grass. Polished pieces of granite form benches and sculptural elements that relate to the building columns.

雕塑庭院引人注目地坐落在西南角，通过在其对面的街角运用一组相似的艺术品作为街景，雕塑庭院与其场地的环境完美地融合。一座由当地艺术家制作、就地浇铸而成的雕塑被置于一块粗制的花岗岩厚板上，其周围还种植着薰衣草和日本银草。抛光的花岗岩碎块制成了长椅和建筑圆柱所需的雕塑原件。

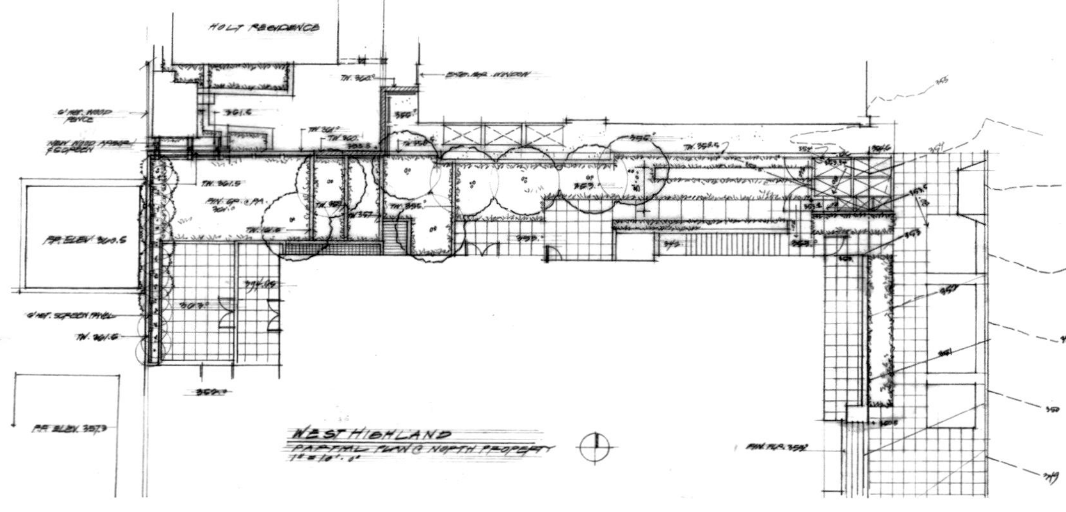

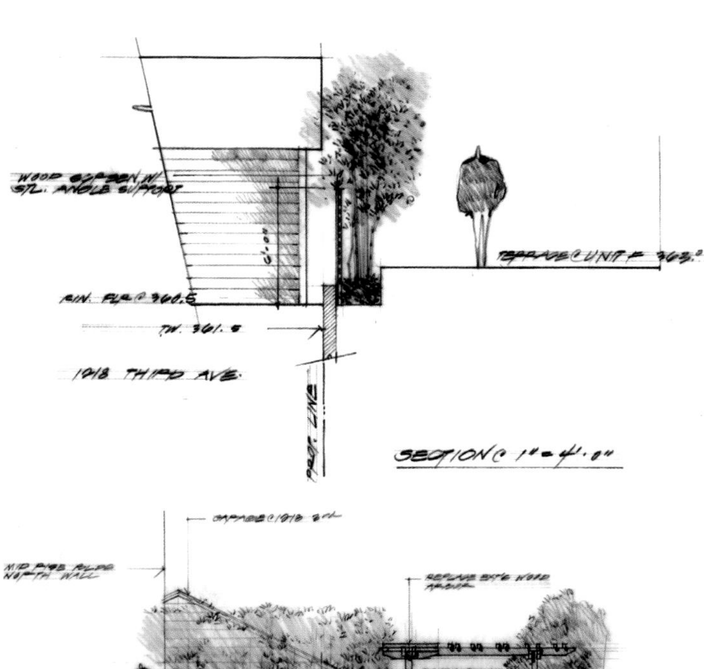
1918 THIRD AVE.
PROP. LINE
SECTION @ 1" = 4'-0"
SECTION @ HOLT TERRACE
1" = 4'-0"

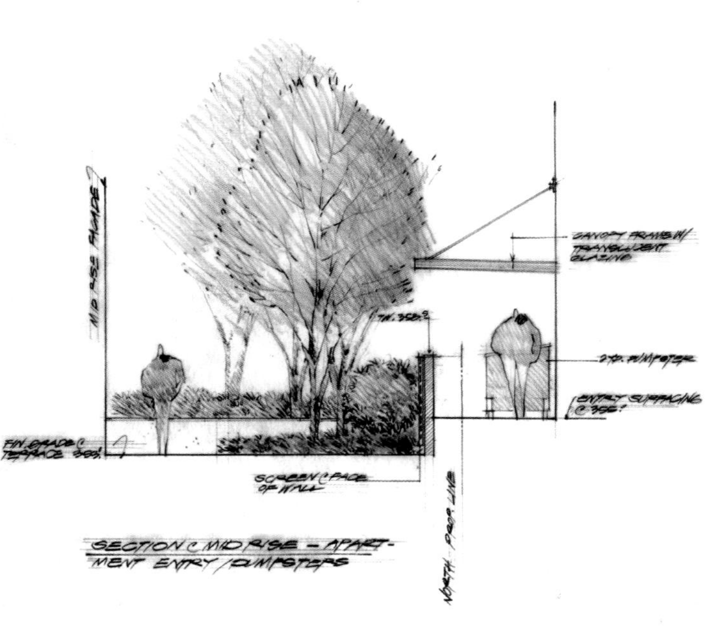
MID RISE FACADE
SCREEN @ FACE OF WALL
NORTH PROP LINE
SECTION @ MID RISE - APARTMENT ENTRY / DUMPSTERS

ISLAND SQUARE
APARTMENTS
APARTMENT AVAILABLE

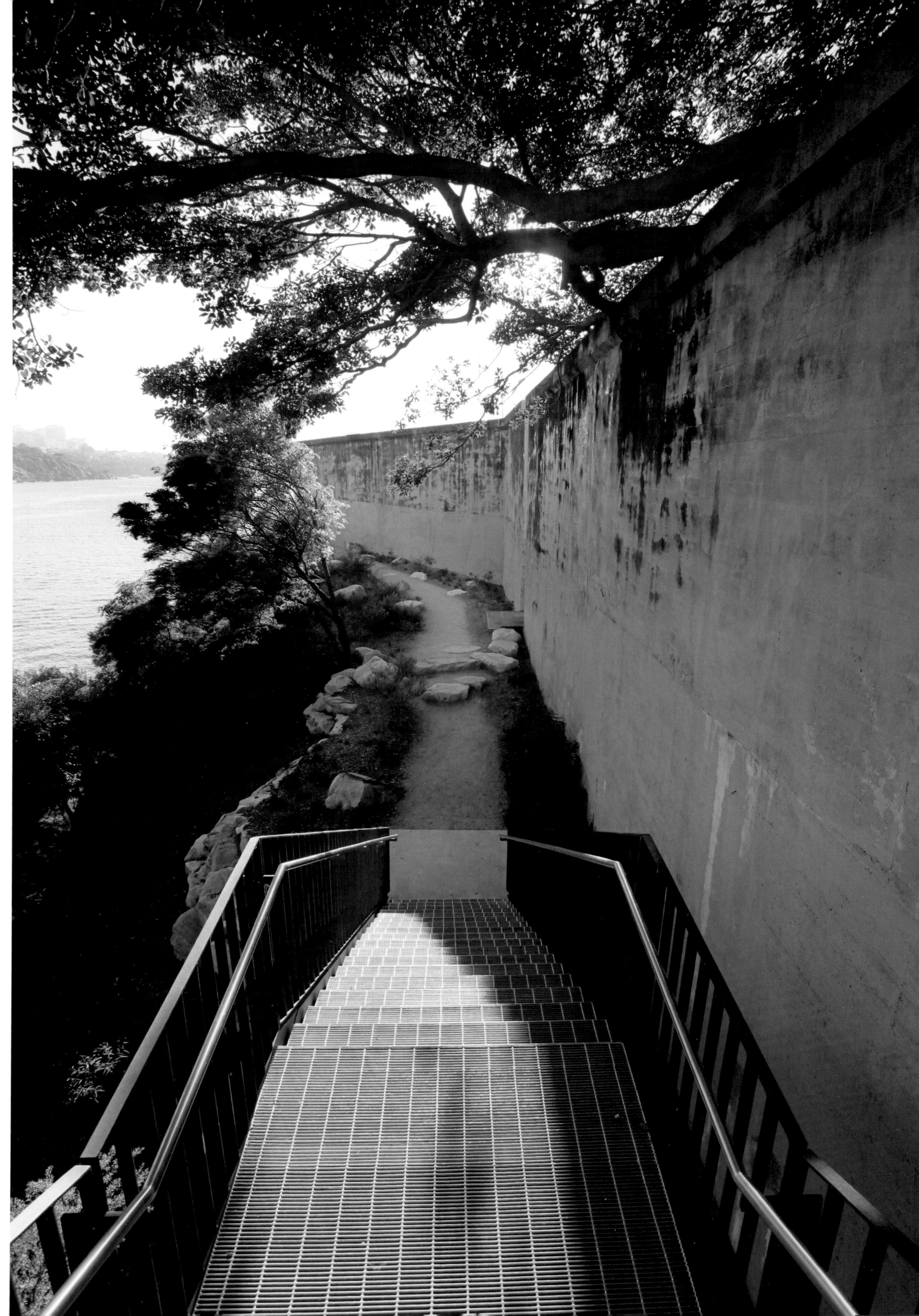

Adrian McGregor
RLA AILA MPIA MAIH
Managing Director

expertise
Landscape Architecture, Urban Design, Master Planning, Project Management
The founding principal of McGregor Coxall has over twenty five years international experience working, teaching and writing about landscape architecture, urban design and the environment. He has worked globally completing projects spanning six nations. Adrian's expertise lies in combining development feasibility, culture and ecology with a passion for design, to create sustainable places in the built and natural environments. His design and mediation skills have been successfully applied to many complex projects bringing communities, authorities and developers together. Adrian has lectured and written many articles on landscape architecture, cities and the environment.

QUALIFICATIONS
Bachelor of Landscape Architecture,
University of Canberra, 1988
Certificate Horticulture, Gold Coast
Tafe 1994

MCGREGOR COXALL

Ballast Point Park

Ballast Point 公园

LOCATION：Sydney，Australia
项目地点：澳大利亚 悉尼

AREA：28,000 m²
面积：28 000 平方米

COMPLETION DATE：2009
完成时间：2009 年

TEAM：Philip Coxall，Adrian McGregor，Christian Borchert，Jeremy Gill，Kristin Spradbrow
团队：Philip Coxall，Adrian McGregor，Christian Borchert，Jeremy Gill，Kristin Spradbrow

DESIGN COMPANY：McGregor Coxall
设计公司：McGregor Coxall

Ballast Point Park

Ballast Point Park is a stunning new harbourside destination, delivered by Sydney Harbour Foreshore Authority on behalf of the New South Wales Government and designed by McGregor Coxall. The park is a result of community action that stopped development of the site for residential development and returned the land to people of Sydney as a parkland. Ballast Point Park was opened to community in July 2009. This project involved McGregor Coxall leading a multidiscipline team in developing the design for this 28,000 m^2 new public park on the former site of a Caltex oil storage and grease manufacturing plant on Sydney Harbour.

Ballast Point 公园是让人叹为观止的海滨胜地。该公园由悉尼海港局代新南威尔士州政府所建，由 McGregor Coxall 设计。先前曾发生一起社会运动，要求将土地归还给悉尼人民，作为公共的公园用地，停止住宅区的扩张。Ballast Point 公园正是这场运动的结晶，它于 2009 年 7 月正式对游人开放。McGregor Coxall 带领一个多领域团队，共同完成这个占地 28 000 平方米的公园设计。该公园选址在悉尼海港原德士古储油仓库和油脂生产工厂的旧址上。

The design uses world-leading sustainability principles to minimize the project's carbon footprint and ecologically rehabilitate the site. The design reconciles the layers of history with forward-looking new technologies to create a regionally significant urban park. The environmental approach is further underpinned by site-wide stormwater biofiltration, recycled materials, and wind turbines designed for on-site energy production.

本设计遵循世界先进的可持续原则，将项目的碳排放量降到最低，恢复该地区的生态状况。本设计融历史特点与前瞻的新技术于一体，力求创造一个非凡的区域城市公园。公园四周的雨水过滤措施、可循环材料的使用及风力涡轮机发电等方式将进一步加强公园的环境措施。

The design challenges our perception of materials and their use. Dominant new terrace walls sit atop the sandstone cliffs but these walls are not made of precious sandstone excavated from another site,

rather from the rubble of our past. What once was called rubbish is now called beauty. It is the new ballast. But it is more than this at play: It is the total composition of the recycled rubber filled cages, off set with concrete coping panels topped with fine grain railing, which allow these walls to sit confidently at the portal to the inner harbour.

本设计冲击了我们对原料应用方式的原有概念。引人注目的砌在砂岩峭壁上的新式扶梯挡土墙并不是由从别处挖掘的珍贵砂岩制成，而是取材自过去的碎石。这正是一种变废为宝的做法，它们重新履行了铺路石的使命。但是它们的意义并未止步于此：循环使用的碎石填满笼子，然后注入水泥，填满缝隙，在碎石板的顶部铺上细沙，这些步骤的共同作用让这些墙体牢牢地屹立在通往内部港口的大门周围。

8 vertical axis wind turbines and an extract from a Les Murray poem, carved into recycled tank panels, form a sculptural re-interpretation of the site's former largest storage tank. The wind turbines symbolise the future, a step away from our fossil-fuelled past towards more sustainable and renewable energy forms.

8个纵轴风力涡轮机，可循环的再生槽板，这些槽板上刻有莱斯·穆瑞诗中的一个句子，共同对这个场地原有的大型储油罐做出了雕塑性的阐述。风力涡轮机象征着未来，是远离原有的石油燃料，趋于可持续能源形式的一步。

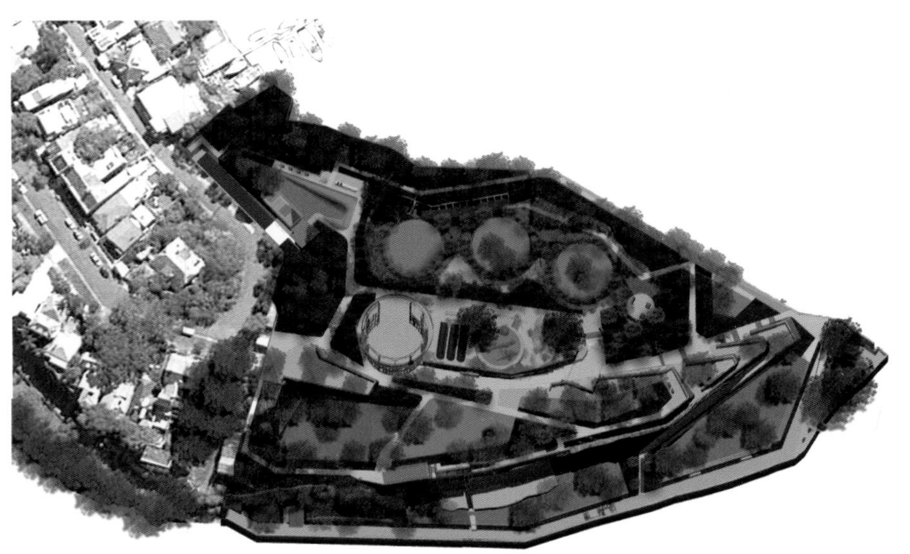

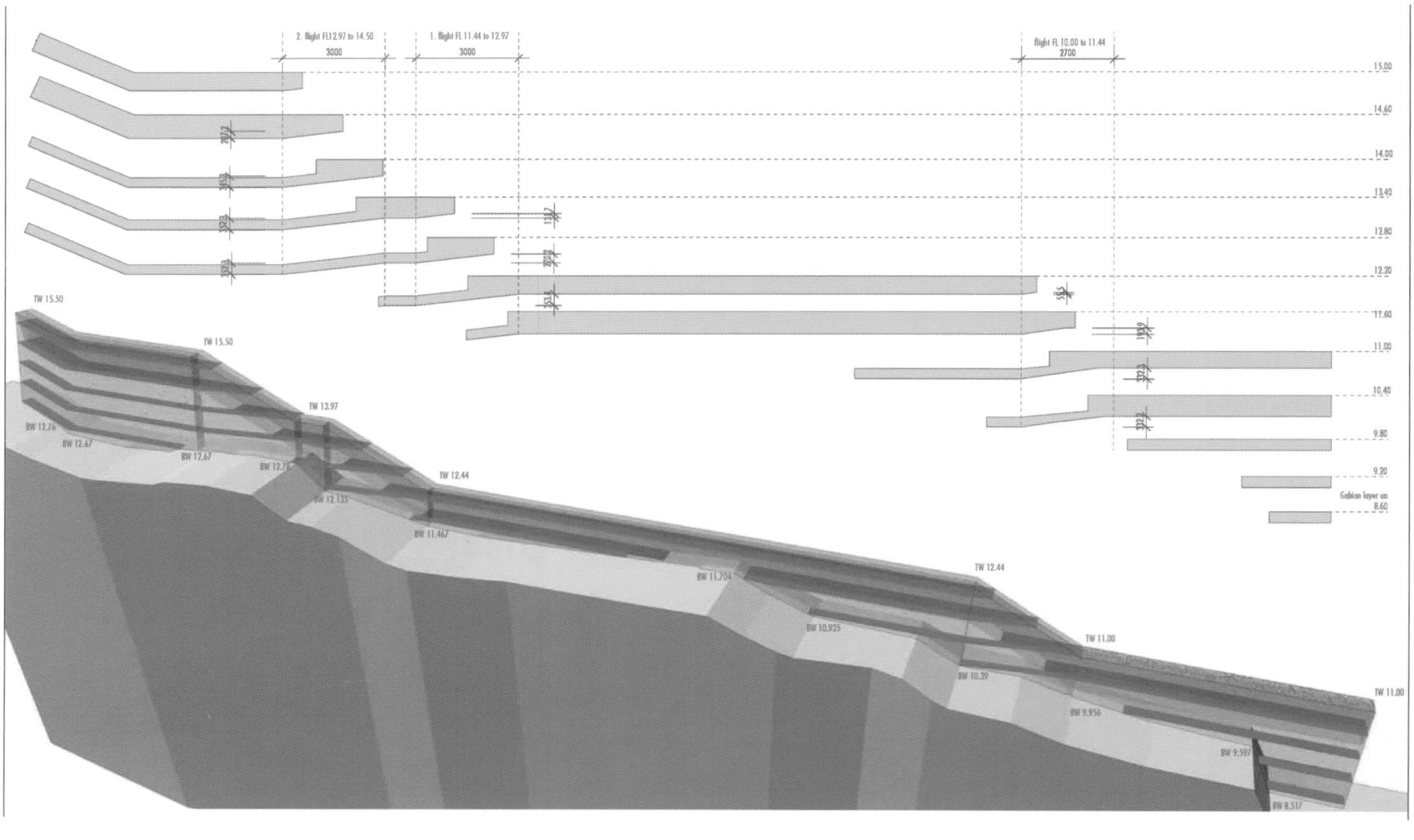

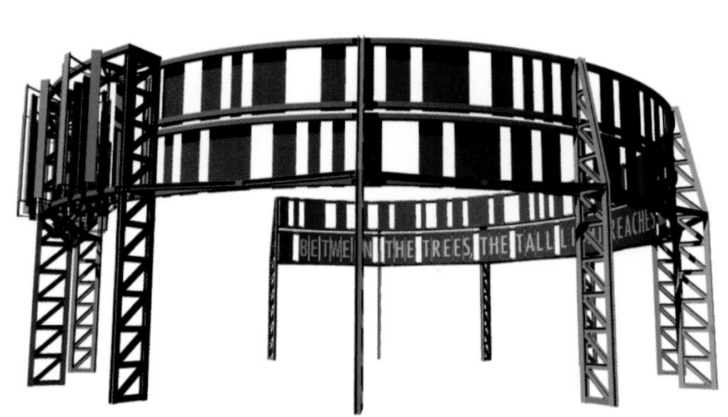

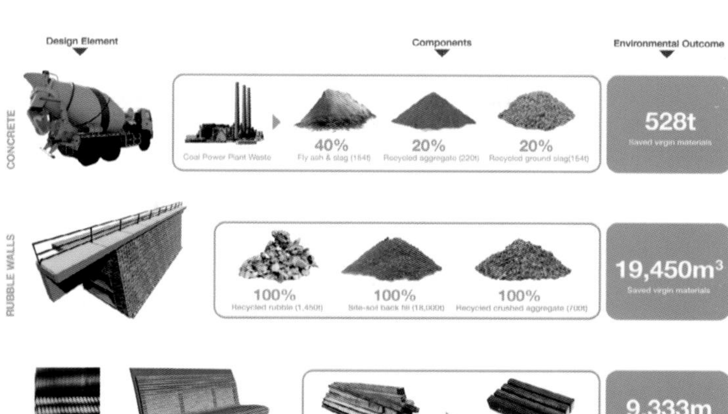
Design Element
Components
Environmental Outcome
CONCRETE
Coal Power Plant Waste
40%
Fly ash & slag (154t)
20%
Recycled aggregate (220t)
20%
Recycled ground slag(154t)
528t
Saved virgin materials
RUBBLE WALLS
100%
Recycled rubble (1,450t)
100%
Site-soil back fill (18,000t)
100%
Recycled crushed aggregate (700t)
19,450m³
Saved virgin materials
TIMBER
Jarrah hardwood
100%
Recycled 30mm x 18mm timber
9,333m
Saved virgin materials
(Linear metres)

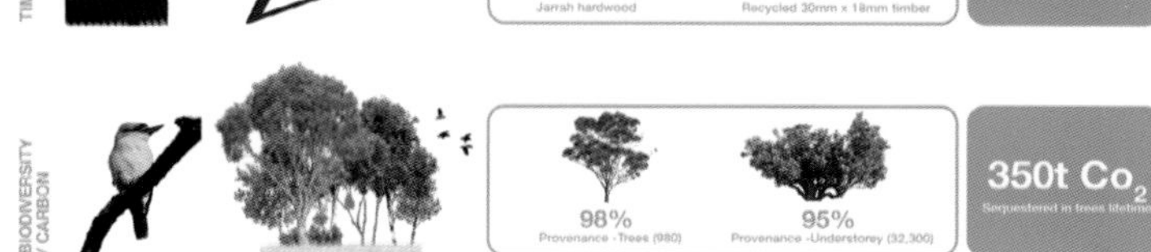
BIODIVERSITY / CARBON
98%
Provenance -Trees (980)
95%
Provenance -Understorey (32,300)
350t Co₂
Sequestered in trees lifetime

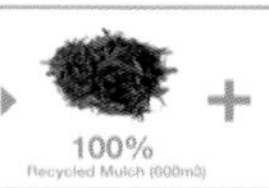

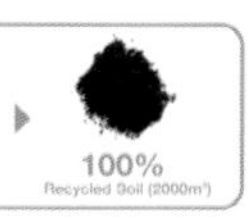

SOIL / MULCH
Used Timber Pallet
100%
Recycled Mulch (600m3)
Organic Waste
100%
Recycled Soil (2000m³)
2,600m³
Saved virgin materials

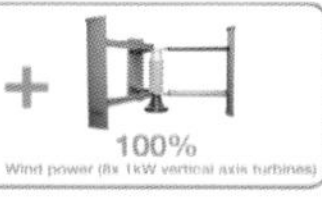

LUBE RING
Demolished 'tank 101'
NGOES
25%
Recycled tank skin
100%
Wind power (8x 1kW vertical axis turbines)
8kW
Wind energy potential

Chang Gung Hospital

长庚医院

LOCATION：Taipei，China
项目地点：中国 台北

AREA：175,000 m^2
面积：175 000 平方米

PHOTOGRAPHER：Cheng Chin Ming
摄影师：Cheng Chin Ming

TEAM：mcgregor+partners（landscape architects urban planners）Ricky Lui and Associates（architects）
团队：mcgregor+partners（landscape architects urban planners）Ricky Lui and Associates（architects）

DESIGN COMPANY：McGregor Coxall
设计公司：McGregor Coxall

Chang Gung Hospital

Chang Gung Hospital at the time of the proposed completion was to be the largest hospital in Asia. It was developed on a 175,000 m^2 site by Formosa Plastics, owners of a number of existing hospitals in Taiwan. McGregor and partners were commissioned to undertake the planning, design, documentation and site review for the projects entire external works. The hospital facility has 3,000 beds that serve general day surgery and a large percentage of patients who stay for periods longer than three months.

长庚医院，这个由台湾塑胶工业股份有限公司所有，占地17.5万平方米的医院是完成本设计时亚洲最大的医院。台湾塑胶在台湾还拥有许多家医院。McGregor和他的同事们共同完成该项目的整个外部工程的计划、设计、搜集资料和场地勘察。医院的3 000张床位可提供普通的日间手术，也能满足大部分住院3个月以上的病人的需求。

The site program was conceived from the idea that physical and psychological experience of a landscape environment is important in the healing process of patients. Design components include extensive roof gardens, an outdoor cafè, interactive water features, walking trails, sun dial, tea gardens and outdoor art/craft workshop plazas. Storm water from the building is channeled to a thirty-meter long cascading viaduct into an elliptical wetland detention pond for recycling as irrigation.

项目的选址来自于对景观环境的生理与心理体验在病人的治疗过程中起着很重要的作用的想法。设计元素包括开阔的屋顶花园。露天咖啡馆、交互式的滨水元素、步行道、日晷仪、茶室、室外艺术／工艺作坊广场。建筑物接收的雨水可沿着梯形的高架桥通向椭圆形的湿地蓄水池，可循环做灌溉之用。

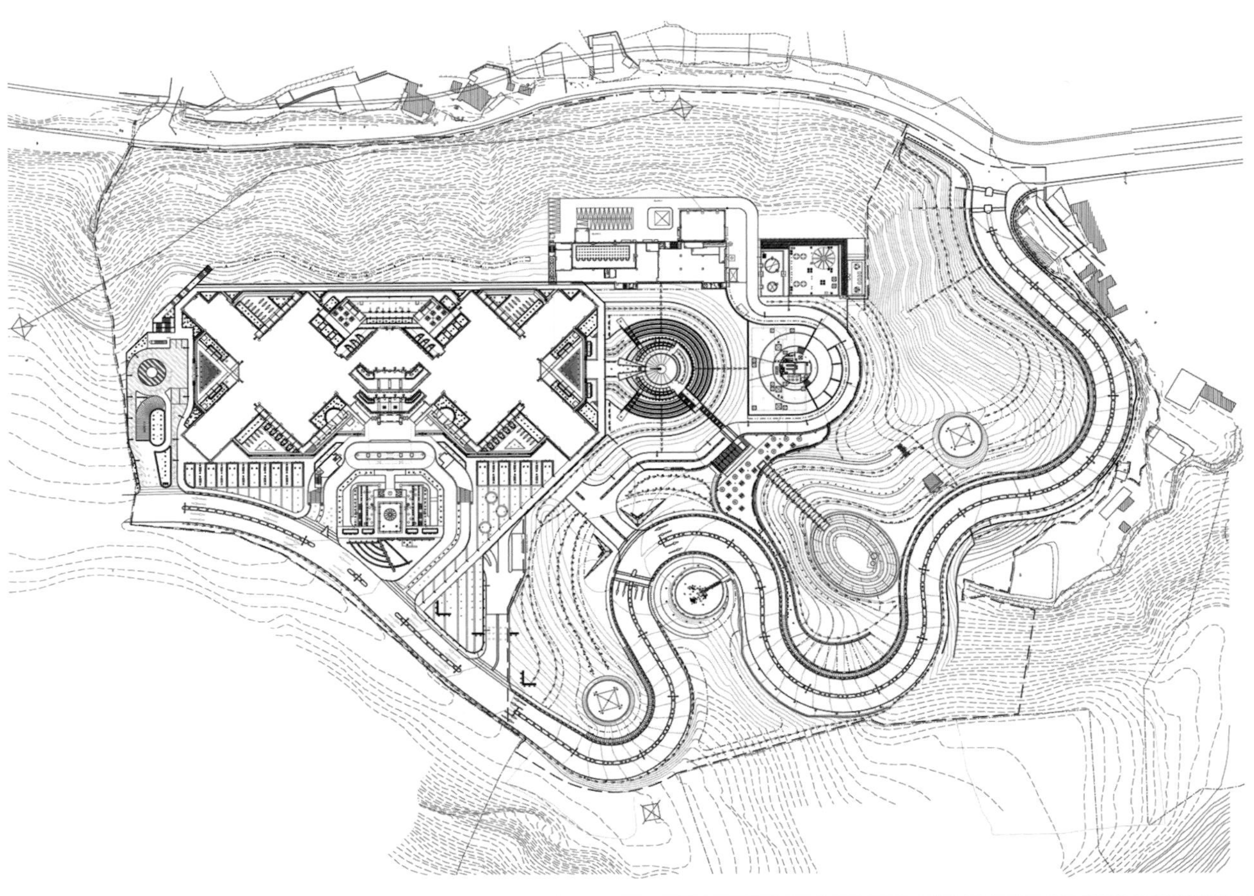

Former BP Site Public Parkland

前 BP 厂址公共公园

LOCATION：Sydney，Australia
项目地点：澳大利亚 悉尼

AREA：25,000 m^2
面积：25 000 平方米

PHOTOGRAPHER：Brett Boardman
摄影师：Brett Boardman

AWARDS：
the National Project Awards for Design（AILA 2006），Design Excellence Award for Landscape and Ecologically Sustainable Development（North Sydney Council 2006），Mcgregor+partners also won the Overall Award for Excellence（AILA 2005）and the Design Excellence Award（AILA 2005）.
奖项：
国家设计项目奖（AILA 2006），景观和生态可持续发展设计优秀奖（北悉尼委员会 2006），Mcgregor+partners 还获得了总体优秀奖（AILA 2005）和优秀设计奖（AILA 2005）。

LANDSCAPE ARCHITECT：Mcgregor + Partners
景观设计师：Mcgregor + Partners

DESIGN COMPANY：McGregor Coxall
设计公司：McGregor Coxall

Former BP Site Public Parkland

Located on Waverton Peninsula, the site is the first of a series of waterfront areas in North Sydney to be transformed from industrial depots into public parklands. The new 25,000 m^2 tare harbourside park is a result of the New South Wales Government's decision in 1997, to convert Waverton's three waterfront industrial sites to public parkland and reject their sale for residential development.

公共公园在 Waverton 半岛上，是北悉尼滨水区第一个由工业石油储放地转换为公共空间的实例。1997 年，新南威尔士政府决定兴建一个占地 2.5 万平方米的海岸休闲公园，将半岛三面滨水工业基地转为公共用地，而不做房地产开发之用。

The former BP site once housed 31 storage tanks, offices and massive concrete perimeter bung walls to prevent oil spills reaching the harbour. The design acknowledges the site's former use through the restrained composition of simple, yet robust structures. The new design celebrates the site's industrial heritage and harbour location with a series of open spaces, wetlands and spectacular viewing decks that embrace the dramatic, semi-circular sandstone cliff cuttings where the oil tanks formally stood. A combination of concrete and metal staircases wrap around the cliffs and project over the water sensitive, wildlife-attracting ecosystem found below.

前 BP 基地拥有 31 个石油存储槽，办公场所和用于防止石油泄漏污染到海港的庞大混凝土墙体结构。设计通过对简单而结实的结构的限制性合成，体现出这里曾是工业用地的特色。新的设计利用一系列的公共空间、湿地和观赏平台表现工业遗迹和其海港的地理位置。在观赏平台上可以看到先前放置储油罐的位置已被改造成了醒目的半圆形砂石铺面岩壁。钢筋混凝土的楼梯沿岩壁环绕而上，引向一片美丽的生态水岸，向下望便可看到适宜野生动物栖息的家园。

Pimelea Play Grounds Western Sydney Parklands

Pimelea 游乐园 西悉尼风景区

LOCATION：Sydney，Australia
项目地点：澳大利亚 悉尼

AREA：20,000m²
面积：20 000 平方米

DESIGN COMPANY：McGregor Coxall
设计公司：McGregor Coxall

Pimelea Play Grounds Western Sydney Parklands

The original Pimelea parklands was constructed prior to the 2000 Sydney Olympics. The Western Sydney Parklands Trust engaged McGregor Coxall to revitalize and extend the parks' facilities by first undertaking a master plan and then designing and documenting a portion of the master plan. This included an upgrade and extension to the existing toilet block, BBQ and picnic facilities, along with an extensive redevelopment of the children's play area. The works also developed a design for a new toilet block and shade elements.

A sensitive response to the rural nature of the site, underpinned by a strong sustainable strategy, drove the site's redesign. Power for the site is generated by solar panels, toilet flushing utilises dam water, all grey water is reused for irrigation and recycled materials are used where possible. The play area extends this concept by introducing recycled water through a children's play pump and water course system within a unique play experience that both excites and delights.

原 Pimelea 风景区建于 2000 年悉尼奥运会前。西悉尼风景区基金委任 McGregor Coxall 对其进行整修和扩建，首先呈交设计蓝图，然后完成整个蓝图局部的设计和编制。整个蓝图包括扩建现有的卫生间模块、改善野外烧烤和野餐设施以及扩建和重新设计儿童游戏区。另外，一个新卫生间模块和绿荫区域也在蓝图构建之中。对景区自然风景的敏锐呼应和对可持续战略的竭力坚持，让这个新设计应运而生。太阳能电池板用于发电，大坝里的水用来冲厕，洗盥水用来灌溉，可循环使用材料物尽其用。在儿童游戏区，孩子们在游戏中能亲自体验水泵和水循环系统，这是寓教于乐，也是循环理念的延伸。

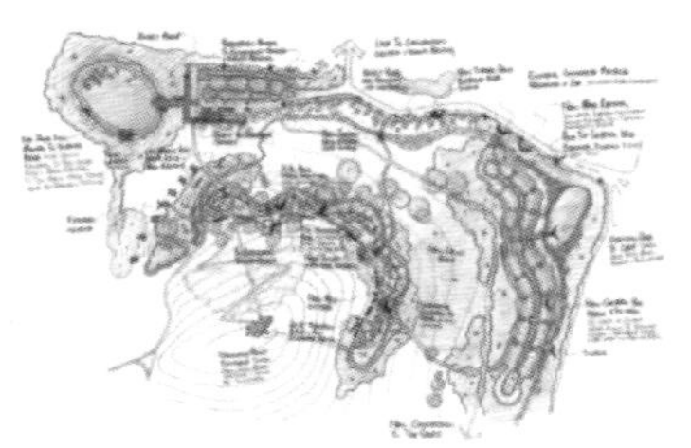

The New Australian Gardens at the NGA Canberra

堪培拉 NGA 新澳大利亚花园

LOCATION：Canberra，Australia
项目地点：澳大利亚 堪培拉，

AREA：36,000 m^2
面积：36 000 平方米

COMPLETION DATE：2010
完成时间：2010 年

PHOTOGRAPHER：Christian Borchert，Simon Grimmett
摄影师：Christian Borchert，Simon Grimmett

The New Australian Gardens at the NGA Canberra

DESIGNER：
Adrian McGregor，Christian Borchert
设计师：
Adrian McGregor，Christian Borchert

TEAM：Kristina Frizen，Georg Petzold
团队：Kristina Frizen，Georg Petzold

DESIGN COMPANY：McGregor Coxall
设计公司：McGregor Coxall

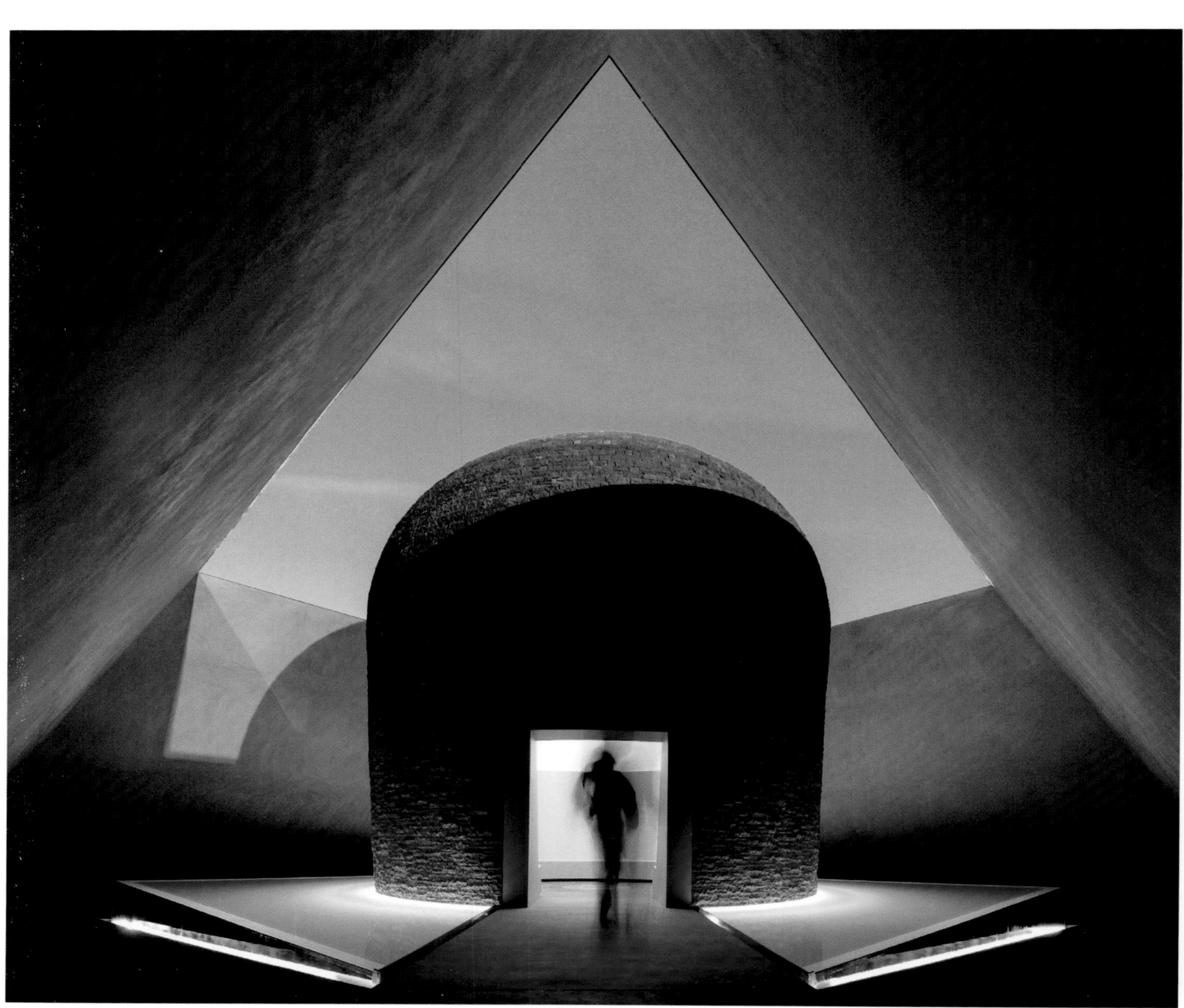

The National Gallery of Australia and its surrounding sculpture garden were completed in 1982 and are located in the arts and civic campus of Australia's National Capital, Canberra. Designed by Edwards Madigan Torzillo and Briggs in the late 20th century "brutalist" architectural style, the Gallery building is characterised by angular masses and raw concrete surfaces. The Gallery is surrounded by a seminal Sculpture Garden planted with Australian native plants designed by the landscape architectural office of Harry Howard and Associates. Both are protected by extensive Conservation Management Plans that recognise the heritage significance of the precinct.

澳大利亚国家美术馆及其周围的雕塑公园坐落于澳大利亚首都堪培拉的城市艺术中心，于 1982 年竣工。其于 20 世纪末由野兽派风格设计师 Edwards Madigan Torzillo 和 Briggs 设计，大量的斗角和粗糙的水泥表面使美术馆大楼颇具特色。美术馆周围环绕着一个半圆形的雕塑花园，由哈里·霍华德及其合伙人景观建筑事务所设计，里面栽种着澳大利亚本土植物。

Rather than address the street, the entrance to the building was designed to meet an elevated walkway intended to connect all the buildings in the precinct. As a result the entrance was hidden and difficult for pedestrians to access from the street. The Gallery quickly expanded it's collection and now has more than 120,000 works. Many of these significant works were not displayed due to space limitations.

建筑物入口不是为了标明街道地址，而是设计成与一条升高的通道交汇，以便从此可通往中心区的所有建筑。结果是入口过于隐蔽，行人从街上走过来非常不便。美术馆的藏品与日俱增，现已超过 120 000 件，由于空间有限，很多重要藏品都不能展出。

In 2005 the NGA embarked on a major refurbishment plan that included a new street entrance, function room, expansion of the indigenous galleries and creation of the Australian Garden. The project also included a major skyspace sculpture titled "*Within Without*" by American artist James Turrell.

2005 年，NGA 提出了一项包括新建入口、功能室、扩建原有美术馆和兴建澳大利亚花园的重要翻修项目。该项目还包括一个由美国设计师詹姆斯·特雷尔所设计的“其内其外”摩天雕塑。

Influenced by the same book that Madigan referenced, "*Space, Time and Architecture*" by Siegfried Giedion, McGregor Coxall ensured the new landscape works embraced the geometric design principles of the

Madigan design by adopting the Golden Mean to proportion new elements. Extending the triangular grid of theoriginalbuilding created a framework for the location and arrangement of significant design elements such as pathways, bridges, walls and water elements.

Siegfried Giedion 所著的《空间时间与建筑》是 Madigan 的设计来源，也对 McGregor Coxall 深有启发。McGregor Coxall 将黄金分割应用于新元素的设计中，以保证新景观与 Madigan 的几何设计准则并行不悖。

The building and landscape were conceived at the outset to be tightly integrated so as to present a unified, legible, accessible public face to the NGA. Located on the previous car park, the main garden was designed around retained Eucalyptus trees. Two planar lawns form the main space creating an "inside–outside" room of huge proportions. The lawns were designed to host functions and events such as temporary art exhibitions and garden parties.

美术馆和周围景观的最初设计构想是紧密相连的，展现出 NGA 统一、明晰和包容的公共空间特色。花园的旧址为停车场，所以主体部分围绕存留的桉树而设计。主空间有两方平坦无垠的草地，营造出一种似远似近，似内似外的空间感。草地主要用来承办现代艺术展和公园聚会等活动和项目。

Visually, the centre piece of the garden is a large pond into which the prominent sculpture "*Within Without*" by James Turrell appears to be sunken. Visitors descend down a ramp through the mirrored water surface of the pond to the interior where the monolithic nature of Turrell's work is evident. In the centre of the sky space is a basalt stupa, a simple domed structure set within a water feature. Visitors move through the stupa to the carefully lit viewing chamber, or oculus, which opens to the sky above.

从视觉上看，花园的中心是一个大型的圆形水池，詹姆斯·特雷尔设计的"其内其外"摩天雕塑仿佛沉在水底。游客从斜坡下来，经过水平如镜的水池，直达内部，在那里可饱览特雷尔的擎天作品。在整个空中花园的中间矗立着一座玄武岩的佛塔，是一座建造在水中的有着简洁的圆形穹顶的建筑。游客经由佛塔可以进入灯光通明的赏景厅或赏景圆孔，观赏头上的苍穹。

Martin Rein–Cano was born in Buenos Aires in 1967. He studied art history at Frankfurt University and landscape architecture at the Technical Universities of Hannover and Karlsruhe. He trained in the office of Peter Walker and Martha Schwartz in San Francisco, and has worked with the office of Gabi Kiefer in Berlin. In 1996, the TOPOTEK 1 office was founded. Martin Rein–Cano has taught as a guest professor in Europe and North America. He has lectured internationally at a variety of universities and cultural institutions, and often serves as a member of design competition juries.

Lorenz Dexler was born in Darmstadt in 1968. He studied landscape architecture at Hannover Technical University. After working at the office of Prof. Günther Nagel, Hannover and Wehberg and at Eppinger Schmittke in Hamburg, he began his collaboration with TOPOTEK 1 in 1996. In 1999, Dexler became a managing partner at the studio. He has lectured at a diversity of schools and institutions, and frequently serves on international design juries.

TOPOTEK I GESELLSCHAFT VON LANDSCHAFTS ARCHITEKTEN MBH

OFFICE PROFILE

The task central to our office is the design of urban open spaces. Based on a critical understanding of immanent realities, the search for conceptual approaches leads us to decided statements concerning the urban context. Throughout design, planning and construction we offer solutions for independent new parks, squares, sports-grounds, courtyards and gardens, whose designs answer to contemporary requirements for variability, communication and sensuality. The manifold experiences through a broad spectrum of German and international projects meanwhile capacitate an efficient realization, finely tuned to respective necessities.

BACKGROUND: TOPOTEK I is a landscape architecture studio that specializes in the design and construction of unique urban open spaces. Founded by Martin Rein-Cano in 1996, the studio's roster of German and international projects has ranged in scale from the master plan to the private garden. Each project strives to respond to site conditions and programmatic necessities with a compelling concept, high quality of design and efficient implementation.

HOW WE WORK: In the early stages of a project, the design teams search for conceptual approaches based on a critical understanding of the task. With digital drawing and physical models we test, revise, and communicate the initial design intent.

TOPOTEK I often collaborates with other creative consultants such as artists, lighting designers and video programmers to enrich the experiential potential of a project. In parallel, we work closely with technical consultants such as civil and traffic engineers early in the design process to integrate site solutions with design innovation consistently throughout the project.

All construction drawings are done in-house to the highest professional and environmental standards. As a project is transferred from the design team to the construction team, the project leader for design continues oversight of drawings, specifications and design revisions. This link ensures that the original conceptual intent stays intact throughout implementation.

Friedrich-Ebert Square, Heidelberg

海德堡弗里德里希－艾伯特广场

LOCATION：Heidelberg，Germany
项目地点：德国 海德堡

AREA：3,500 m^2
面积：3 500 平方米

COMPLETION DATE：2008
完成时间：2008 年

DESIGN COMPANY：Topotek 1 Gesellschaft Von Landschaftsarchitekten Mbh
设计公司：Topotek 1 Gesellschaft Von Landschaftsarchitekten Mbh

Friedrich-Ebert Square, Heidelberg

The Friedrich-Ebert square forms a compact urban unit, which is reactiveted by simple materials and language. The revitalization of the town square and market is made possibly by the basic idea to exempt the inner space to realize a simple and functional space layout. The rows of trees along the west and east sides direct the view to the hillside areas of Odenwald forest in the south. To gain an open view into the square, the bus stop is oriented in north-south direction on the southern end of the square. A grey sandstone is used to make a homogeneous surface throughout the site. The sidewalks of the streets are paved with the same material, to give the impression of a connection to the facedes of the buildings. The ramps to and from the parking are positioned at the southern part of the square close to the site. The are made out of asphalt integrated into the color concept.

弗里德里希－艾伯特广场形成一个布局紧密的城市单元，而简约材料和特殊语言的运用为广场重新注入了活力。城市广场和市场之所以能重新散发出活力，主要是基于一个基本思想，就是解放内部空间，以实现简单而实用的空间布局。东西两侧一排排的树木将人们的目光直接引向南面奥登森林的山坡上。为了使广场的视野开阔，公共汽车站呈南北方向布置，位于广场的南端。整个场地使用同一种灰色的石英砂岩，形成均匀的表面。周围街道上的人行道也采用了相同的材料，感觉像是和建筑物的外立面连接在了一起。进出地下停车场的坡道位于广场南面很近的地方，坡道采用沥青表面，其中还融入了色彩的概念。

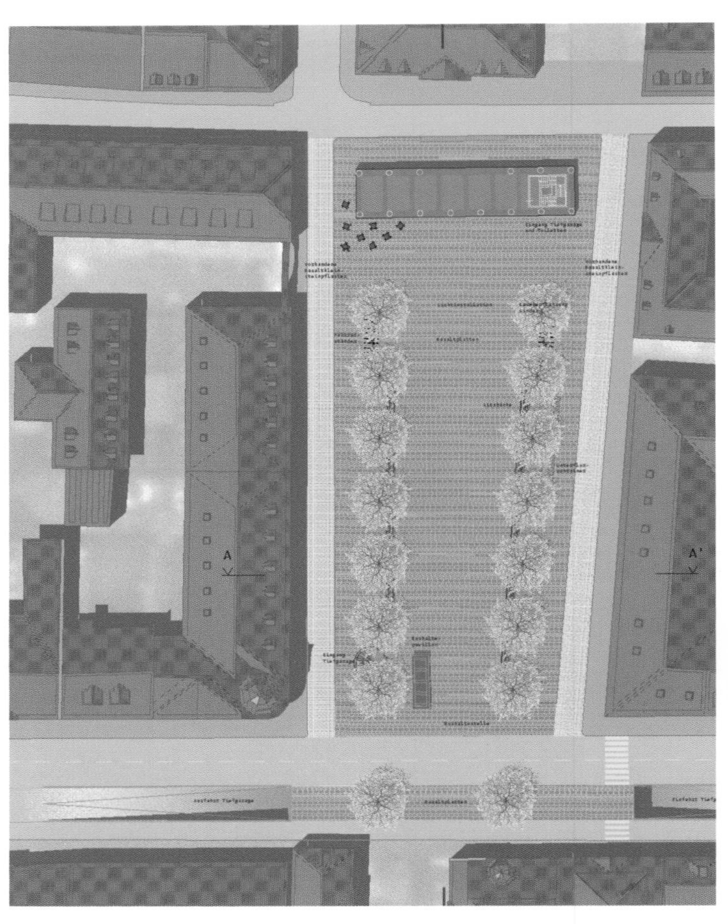

National Garden Show Schwerin 2009

2009 什末林国家花园展

LOCATION：Schwerin，Germany
项目地点：德国 什末林

Size：31，000 m^2
面积：31 000 平方米

COMPLETION DATE：2009
完成时间：2009 年

DESIGN COMPANY：Topotek 1 Gesellschaft Von Landschaftsarchitekten Mbh
设计公司：Topotek 1 Gesellschaft Von Landschaftsarchitekten Mbh

National Garden Show Schwerin 2009

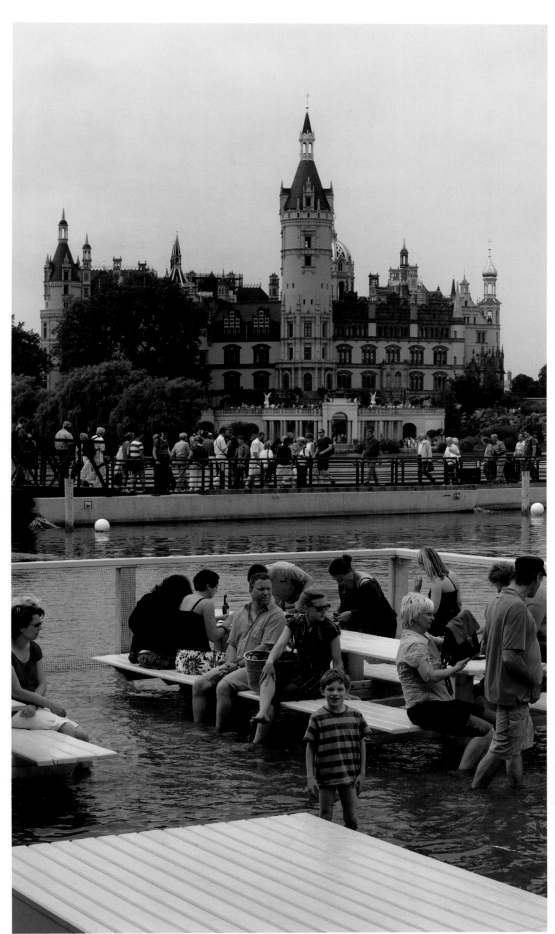

Water is the central issue in the design for the shore garden. A new layout of paths enables a variety of approaches to the water and a multitude of possibilities to explore its qualities. A path serves to reach several stations next to, on, and in the water. The path has a formal language of its own. Its assembled linear stretches form a dynamic motif; and they bend slightly, open up and narrow down again. Along the way plantings of perennials, annuals and riparian vegetation enhance the general water theme of the garden.

"水"是海滨花园设计的中心问题。小路新的布局使人们能够以各种方式接近水面，探究水的特质。一条小路通向水边、水上和水中的各个站台。这条小路有自己正式的语言。它的分支组合在一起，呈线性延伸，形成了一个动态的主题：时而弯曲、时而开阔、时而狭窄。沿着小路生长着一些多年生、一年生和水滨植物，加强了花园"水"的主题。

The visitors reach the shore garden coming from the castle. The path along the waterfront takes them across an estuary of the lake, through reeds and a glade with a playground—all the way to the terrace of the rowing club. Here, overseeing picturesque water—lily fields the terrace affords a panoramic view across the lake and back towards the castle. The path then follows the natural shore for a while, goes under a shady tree canopy, and finally reaches a temporary beach where white sand and palm trees create a surreal piece of maritime landscape. From the beach the path swerves onto the lake and leads to the other side of the lake on a pontoon construction. In the middle of this footbridge a floating water—lounge is moored, accentuating and structuring the stretch of way across the lake. Here, at the bar one can have a cocktail or a cool, refreshing glass of water.

游客可以从城堡来到海滨花园。水边的小路带领着游客穿过湖港、芦苇和一片有运动场的空地——一路来到划船俱乐部的平台上，从平台上可以俯瞰如诗如画的睡莲，一览湖面和后面城堡的全景。然后，小路沿着自然海岸延伸，穿过一片树荫，最终来到临时海滩，海滩上的白砂和棕榈树形成了一幅梦幻般的海景。从这里开始，小路转向湖面，通过一个浮桥到达湖的另一侧。在浮桥的中间有一个漂浮着的水上休息室，突出了湖上通道的延展，在这里，人们可以在酒吧品尝鸡尾酒，或者喝一杯清凉的水。

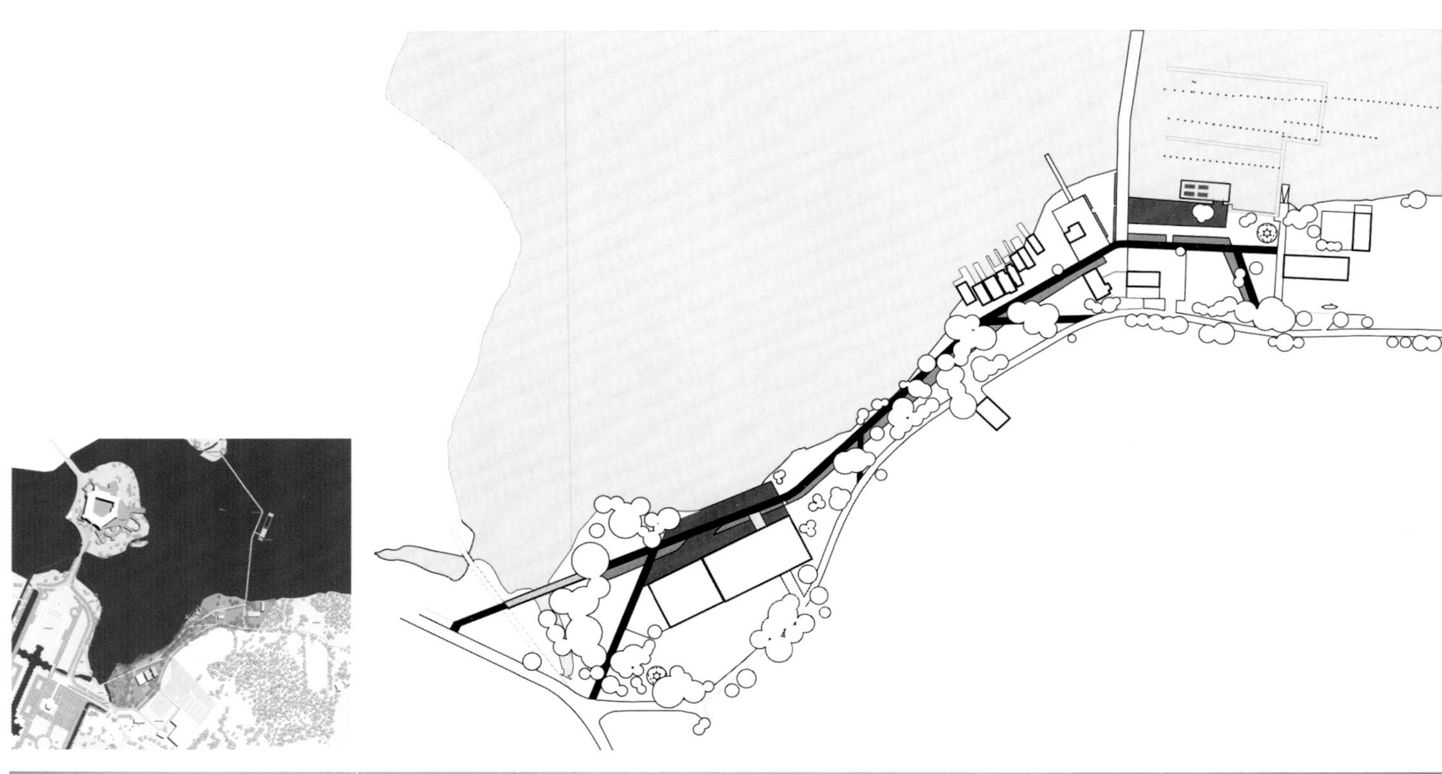

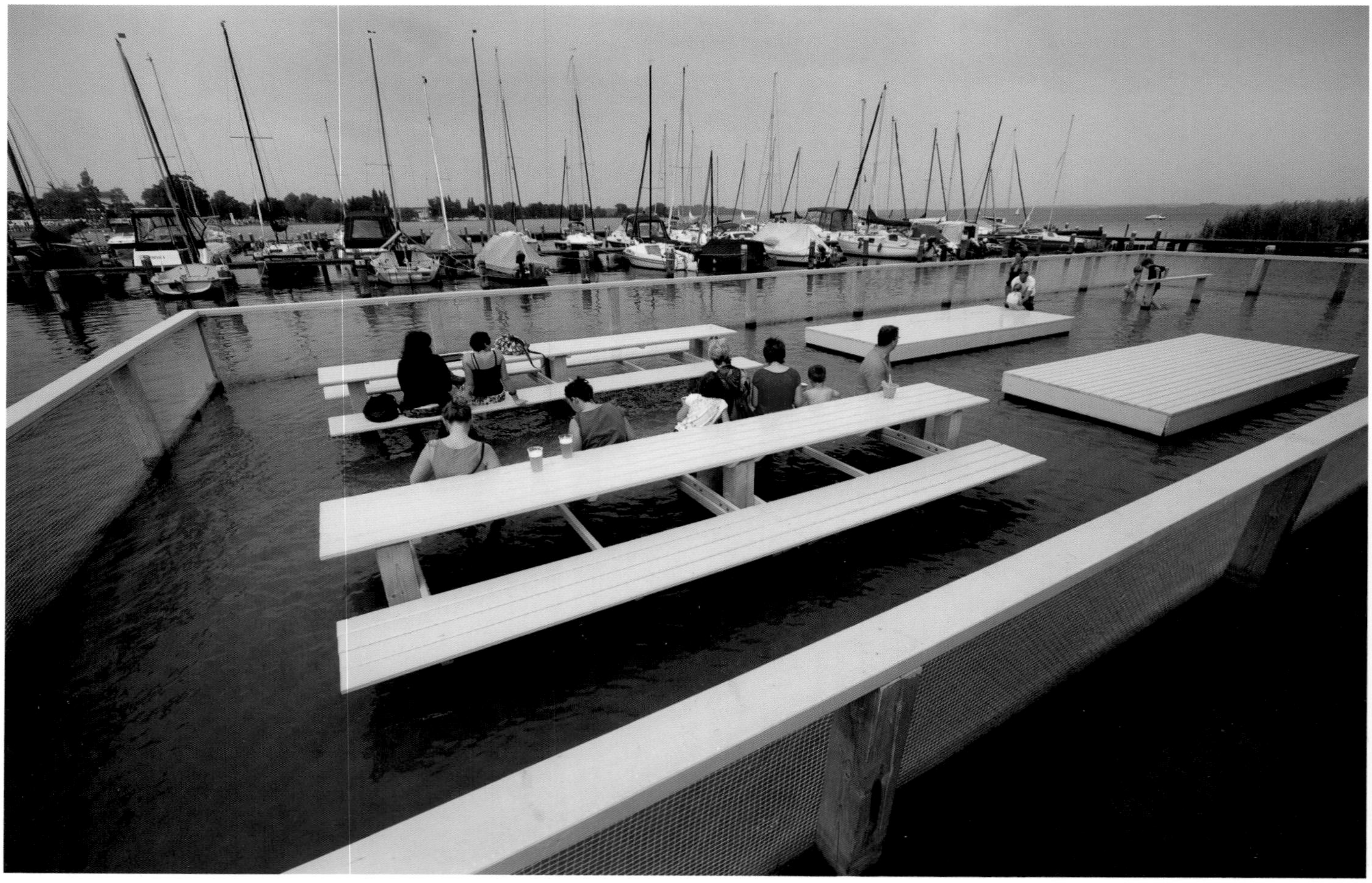

Augsburg Wollmarthof

奥格斯堡 Wollmarthof 住宅

LOCATION：Augsburg，Germany
项目地点：德国 奥格斯堡

AREA：1，600 m^2
面积：1 600 平方米

COMPLETION DATE：2009
完成时间：2009 年

DESIGN COMPANY：Topotek 1 Gesellschaft Von Landschaftsarchitekten Mbh
设计公司：Topotek 1 Gesellschaft Von Landschaftsarchitekten Mbh

Augsburg Wollmarthof

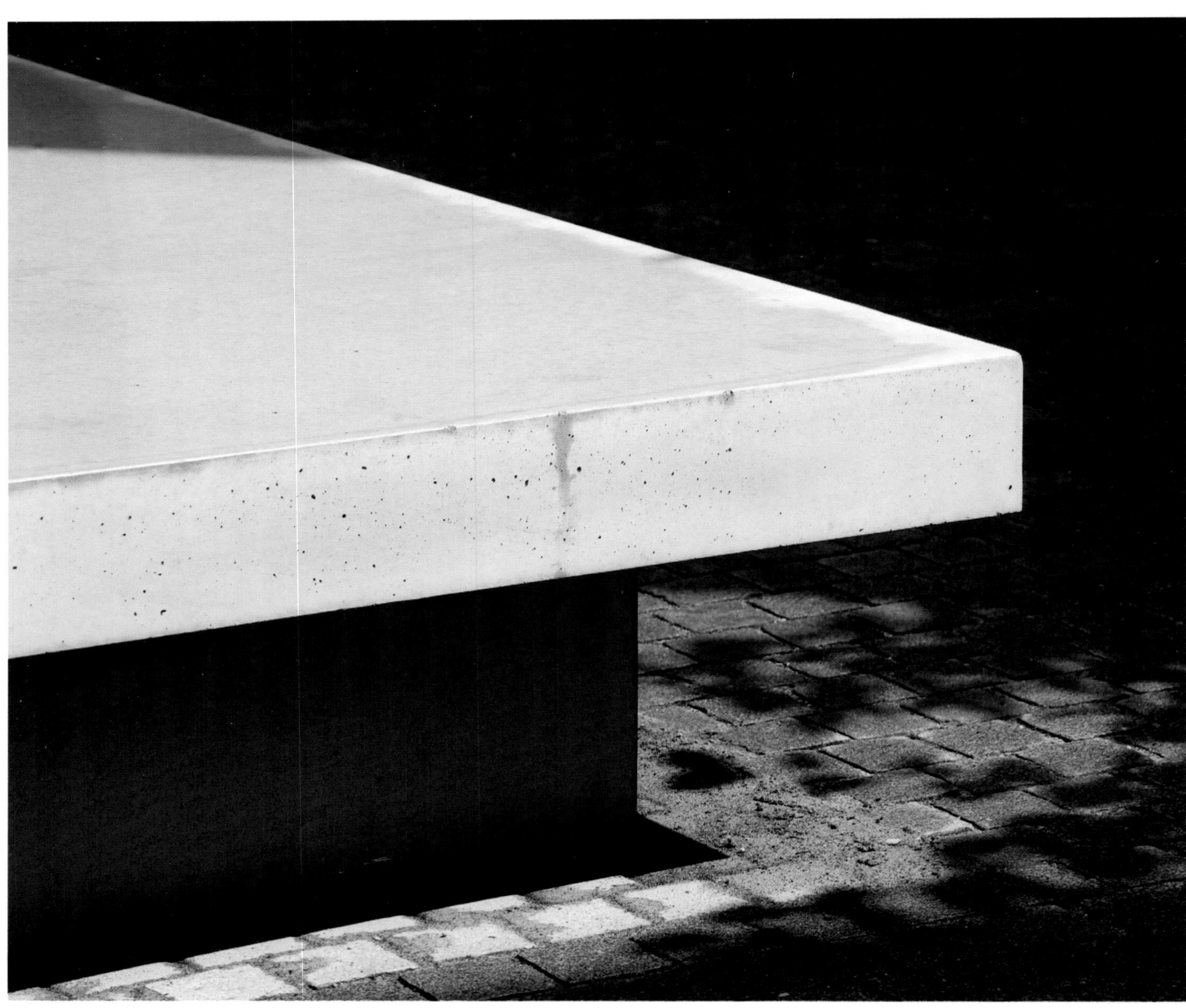

The Wollmarthof is a heterogenic space: Different functions of the buildings exist next to each other without communicating with each other.

Wollmarthof是一个与众不同的空间：建筑的各个功能区都彼此相邻，但是彼此之间又没有联系。

The old cloister and the St Margareth Church are currently the only remaining pieces of a dispersed ensemble.

古老的修道院和圣玛格丽特教堂是目前历史遗留下来的唯一痕迹。

The main idea of our proposal is to make all the historic facades of the buildings visible and to solve the problems which are caused by the public and private use of certain areas, by the creation of an expanded public open square.

方案的中心思想是再现建筑的历史外观，并通过建造一个更大的公共广场解决某些区域在个人使用和公共使用中出现的问题。

An overall pattern emerges with the use of small sized paving stones, which recover the honor of the historic facades.

我们使用较小的铺路石打造出一个总体图案，再现了建筑历史外观的辉煌。

The surfaces of the used paving stones will be cut and assembled into a new mosaic pattern. Missing stones will be replaced by new ordered ones to create the desired image.

对用过的铺路石进行表面切割，然后拼成一个新的图案。丢失的铺路石要重新定购，以便能够打造出要求的图案。

The final image will compose all the different areas to one expanded surface.

最终的图案将把所有的区域都组合成一个展开的表面。

As an answer to the hedge two large, green marble benches will be placed along the front facade of the historic cloister. They provide a place to rest and also underline the entrance situation in front of the historic arcades.

古老修道院的正面设置了两个绿色的大理石长凳，与绿篱相互映衬，不仅提供了休息的地方，而且突出了古老拱廊前面的入口。

Theresienhöhe Railway Cover Munich

慕尼黑 Theresienhöhe 铁路覆盖区

LOCATION：Schwerin，Germany
项目地点：德国 慕尼黑

Size：1,600 m²
面积： 1 600 平方米

COMPLETION DATE：2009
完成时间：2009 年

DESIGN COMPANY：Topotek 1 Gesellschaft Von Landschaftsarchitekten Mbh
设计公司：Topotek 1 Gesellschaft Von Landschaftsarchitekten Mbh

Theresienhöhe Railway Cover Munich

The "open space" project of the former trade fair area Riem, in Munich includes the 300–meter long and 50–meter wide concrete plates above the railway connecting Munich to Rosenheim. The total area will be a new district of Munich offering office and residential buildings, shops, and a school.

原慕尼黑里姆商品交易会的“公共空间”是一片长300米，宽50米的混凝土场地，位于慕尼黑到罗森海姆的铁路上方。这片区域将成为慕尼黑的一个新区，用来建设办公和住宅楼、商店、幼儿园和学校。

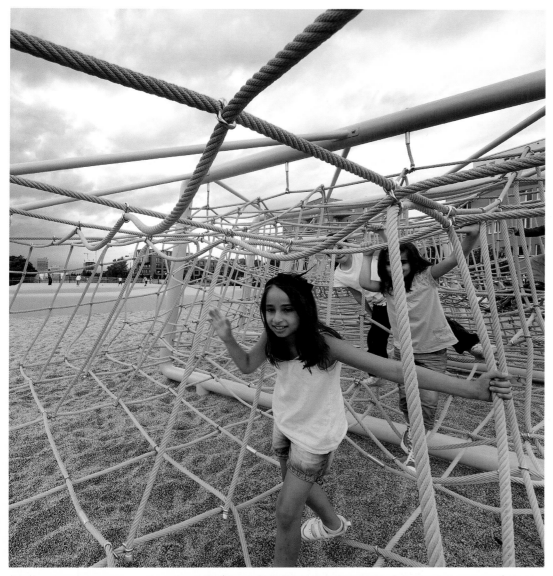

Together with artist Rosemarie Trockel and architect Catherine Venart, the "Bahndeckel" is designed to be an important open space in this new area. The chaste and clear concept consists of an orange and green lane–a multifaceted urban space, connecting three landscape elements and combining different materials. A rubber–paved strip moves from one side to the centre of the stretched out space, and a lawn–strip comes from the other side. In the centre a sand dune rises above the area whose edge is subject to constant change due to permanent usage. The entire space is accessible from all sides. A low wall to sit on surrounds the area, turning into a higher supporting wall along the grass section. Along one side of this wall an over–dimensioned ball–net is spanned. In the background, accentuating pine trees are planted in large pots, providing shade with their crowns.

"Bahndeckel"由艺术家罗斯玛丽·特洛科尔和建筑师凯瑟琳·沃纳特共同设计，成为这个新区的重要公共空间。设计采用朴实、明晰的理念，打造出一条黄绿相间的车道 ，形成一个多方位的城市空间，融合了不同的材料，将三个景观元素联系到一起。一条橡胶路从一侧通向空间的中央，而另一侧则采用草坪。场地的中间有一个沙丘，由于这里长期有人使用，沙丘的边缘不断变化。场地从各个方向都可以进入，四周是可以当做座椅的矮墙，草坪处是高一点的支撑墙；墙的一侧有一张巨大的球网。大花盆中种着一些松树，松树的树冠为人们提供了纳凉之所，成为场地一个突出的背景。

This wide and open space stands in contrast to the dense urban structure of the adjacent buildings. With its many uses the "Bahndeckel" will become a new attractive urban square.

这片开敞的空间与附近建筑形成的密集城市结构形成对比。“Bahndeckel”提供了各种功能，将成为一个新的魅力城市广场。

Sports Facility Heerenschuerli

Heerenschuerli 体育中心

LOCATION：Zurich，Switzerland
项目地点：瑞士 苏黎世

AREA：100,000 m^2
面积：100 000 平方米

COMPLETION DATE：2007
完成时间：2007 年

ARCHITECT：Duerig Architekten AG
建筑师：Duerig Architekten AG

DESIGN COMPANY：Topotek 1 Gesellschaft Von Landschaftsarchitekten Mbh
设计公司：Topotek 1 Gesellschaft Von Landschaftsarchitekten Mbh

Sports Facility Heerenschuerli

The Sports Facility Heerenschuerli, one of the largest sports facilities in Zurich, is located among a nature reserve, a highway crossing, residential housing and large-scale industrial buildings. It will be thoroughly restructured to cater to the changing and expanding demands of sports and leisure activities. A compact, extremely popular sporting world with twelve football fields and one for baseball are integrated with three buildings: an ice-rink, a locker-room building and a workshop. The playing fields are all surrounded by high fences, giving the sports arena its architectural status and urban poignancy. The fences form diverse spatial sequences mediated along a simple path layout. The determinant connecting function of the orthogonal path axes is interwoven with the sensuality of sport and connected with the spatial and functional core of the complex over a series of rows of trees, avenues and accent-setting groves. The overlapping transparency of the fences creates a unique visual dynamism which is further amplified by the materials used in their construction: the two-ply, wire-mesh walls in various green tones produce a moire effect which makes the overlapping of the various spatial layers a part of the mise-en-scene of the relationship between athletes and spectators.

Heerenschuerli 体育中心是苏黎世最大的体育中心之一，位于自然保护区、公路交叉口、住宅建筑和大规模的工业建筑之间。为了满足不断变化和增长的体育休闲活动的需求，Heerenschuerli 体育中心将进行彻底的改造。这里将建成一个布局紧凑、非常受欢迎的体育世界：包括十二个足球场（其中一个用于棒球场）、一个室内溜冰场、一个更衣室和一个工作室。运动场全部用高大的围墙围合，突出了运动场的建筑地位和城市色彩。围墙在一条简单的小路上形成多样化的空间序列。小路的轴线成直角，主要起连接功能，与运动的感性交织在一起，并通过一排排树木、道路和树丛与建筑的空间和功能核心连接起来。围墙采用相互重叠的透明材料，形成一种独特的视觉动感，而在建造过程中使用的材料进一步加强了这种动感：不同绿色的双层金属网墙形成一种莫尔效应，使相互重叠的不同空间层次成为运动员和观众之间关系的一部分。

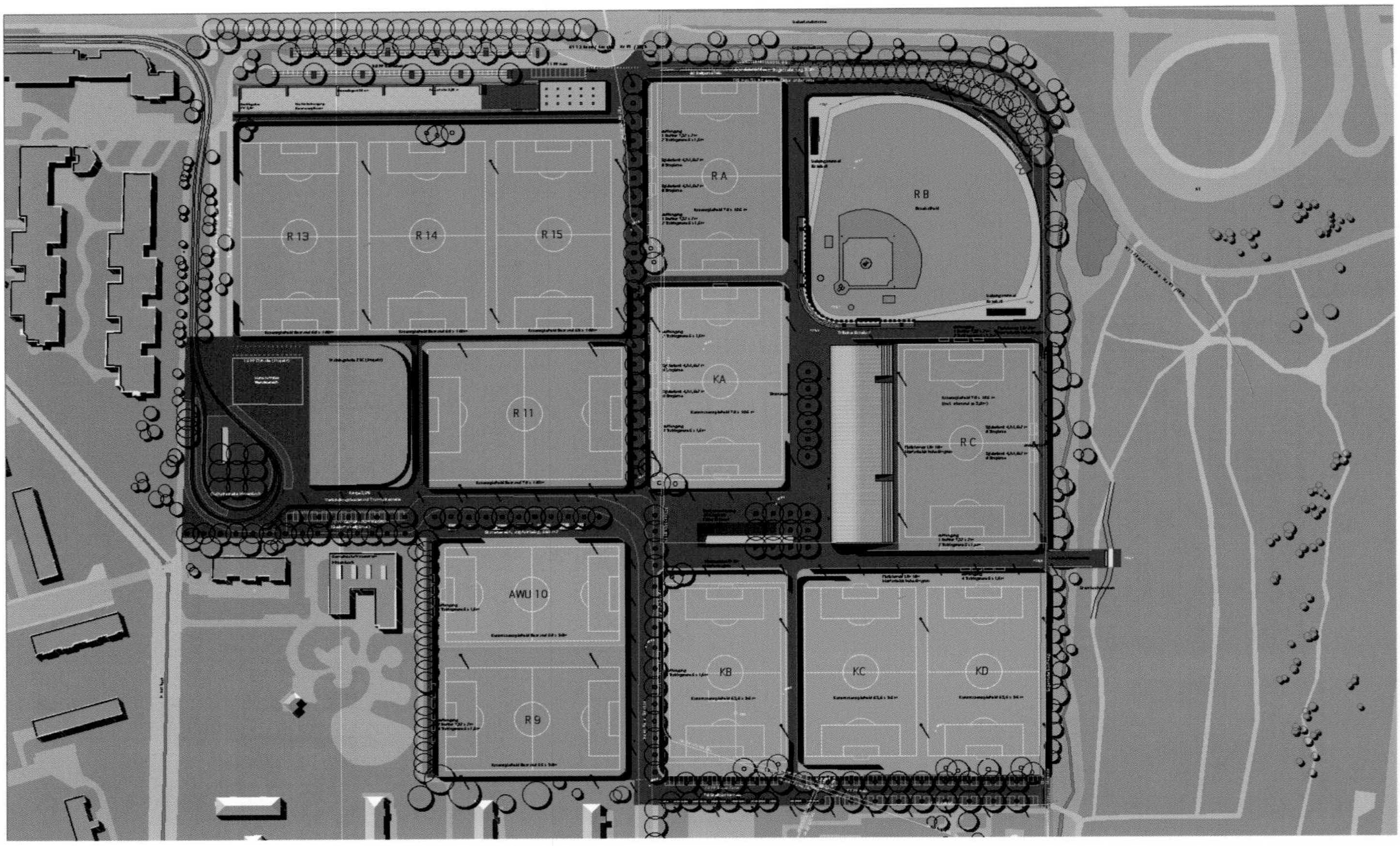

R 13
R 14
R 15
R A
R B
K A
R 11
R C
AWU 10
K B
K C
K D
R 9

Bergannstrasse 71, Berlin

柏林伯格曼大街 71 号

LOCATION：Berlin，Germany
项目地点：德国 柏林

AREA：3,000 m^2
面积：3 000 平方米

COMPLETION DATE：2009
完成时间：2009 年

DESIGN COMPANY：Topotek 1 Gesellschaft Von Landschaftsarchitekten Mbh
设计公司：Topotek 1 Gesellschaft Von Landschaftsarchitekten Mbh

Bergannstrasse 71, Berlin

The former post office at the eastern side of Marheinecke Square in Berlin Kreuzberg was converted into a modern state of the art office and retail building. Within this development one part became high extension of the basement retail area.

原来位于柏林克洛伊慈贝格区 Marheinecke 广场东侧的邮局现在改造成了一个先进的现代化美术和零售大楼。在这次改造中，有一部分变成了地下零售区。

The additional wing and a garden wall are separating the inner garden area from a functional outer car park area. A terraced garden area planted with sculptural multistem gingko trees is connecting the generous terrace with the green courtyard. The trees and the understory planting with evergreen herbaceous grasses achieve a natural antipole to the formal structure of the terraces.

附加的耳房和花园的一面墙把室内花园与室外停车场分开。平台花园中种植了雕塑般的多干银杏树，将宽敞的平台与绿色的庭院联系在一起。树木与树下常绿草本植物与平台正式的结构形成对比。

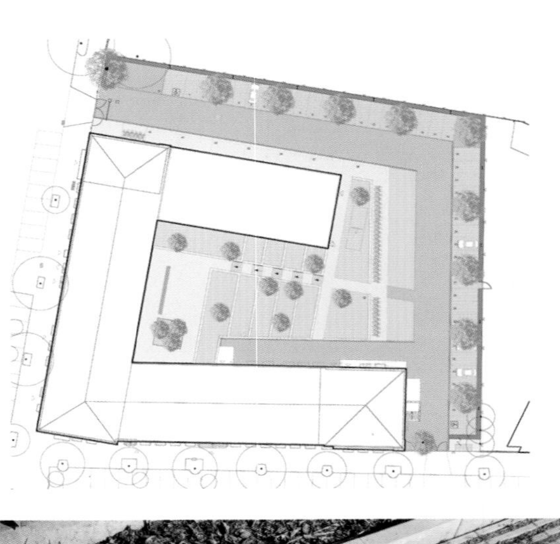

THE Big Dig

大挖掘

LOCATION：xi'an, China
项目地点：中国 西安

AREA：1,000 m^2
面积：1 000 平方米

COMPLETION DATE：2011
完成时间：2011 年

DESIGN COMPANY：Topotek 1 Gesellschaft Von Landschaftsarchitekten Mbh
设计公司：Topotek 1 Gesellschaft Von Landschaftsarchitekten Mbh

THE Big Dig

The idea of going through the Earth and coming out on the other side must be an ancient dream.
A common childhood warning of, "If you keep digging, you'll dig all the way to China" is often an introductory presentation of this dream. The whole idea is a spirit of adventure and whimsy, and the unending curiosity of what can be found on the other side. With a garden designed in Xi'an, China, we couldn't resist: What if we did dig all the way to China?

古代人一定有过这样的梦想：穿过地球，然后从另一面出来。还有一个对儿童的警告与这个梦想也很相似，警告是这样说的："如果你一直挖下去，你就会一路挖到中国。"这个梦想反映出一种探险的精神和奇怪的想法，也是对在另一面会发现什么的好奇。在为中国西安的一个花园进行设计时，我们不禁想到：如果我们真的一路挖到中国，会怎么样呢？

In the garden we have created the result of this overzealous dig: the point that the hole emerges in China. At this hole we capture a precise point where one stands on the edge of one world and the other, wondering what is possible of the other side?

在花园中，我们就进行了这样一次大挖掘：这里就是洞在中国的出口。在这个洞口，我们找到了一个精确的点，一个人站在这一点，世界的一端，想象着另一端会是什么样子？

A deep hole in the ground, close to the shape of a speaker or an ear, is circumscribed by a glass railing to prevent people getting too close to the edge and falling down, but it is also a cinema screen creating a thin line between reality and fiction. Artificial grass covers all surfaces, the hole, the lawn and the bench, and creates a monolithic appearance.

这是地面上的一个深洞，与扬声器或耳朵的形状类似，用玻璃围栏围起来，防止人们由于离洞口太近而掉到洞里面，不过，玻璃围栏也是一个电影院的屏幕，在现实与虚幻之间形成一条细线。人造玻璃覆盖了所有的表面、洞口、草坪和长凳，形成了一个整体的外观。

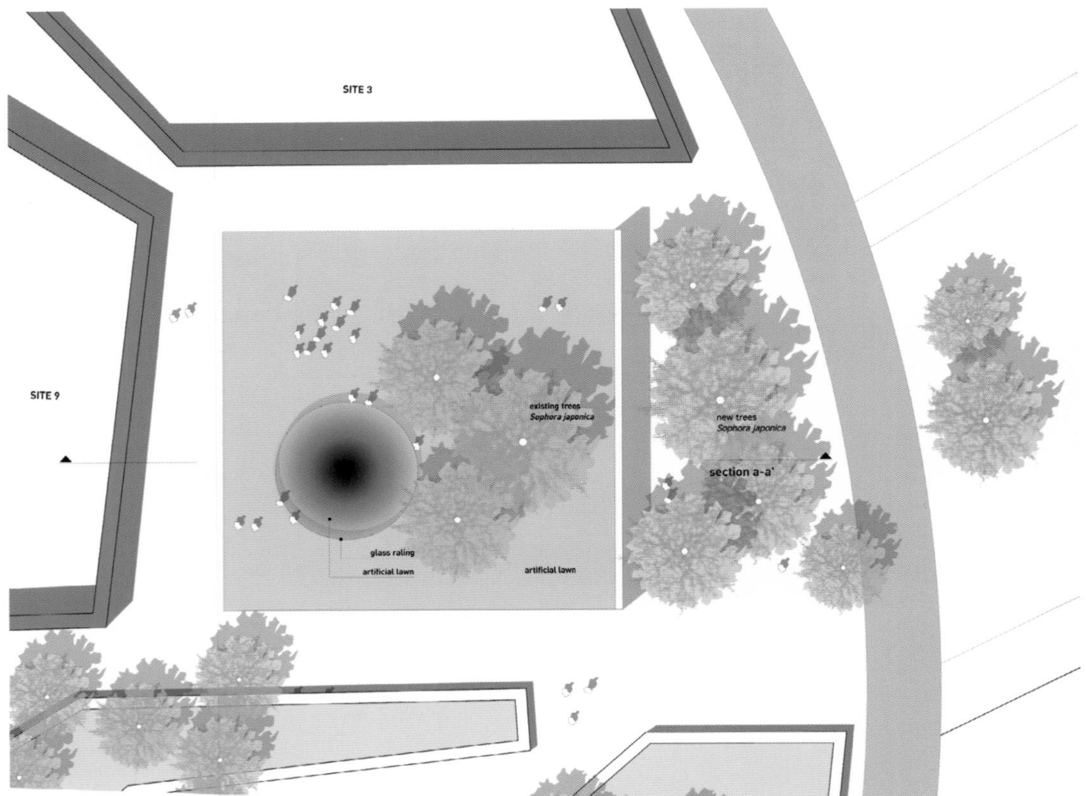

The garden offers a visual and auditive journey in different spheres. When you come close to the edge, a sound installation stages discreetly the ambience of travelling. From the hole in the ground you can hear the other side of the world: the sounds of Berlin Central Station.

在花园中，不同的地点就会有不同的视听感受。来到洞口，一个声音装置就会模拟出旅行的氛围。从地面的洞中你可以倾听世界的另一端柏林中央车站的声音。

This hole through the Earth connects people from the diametrical locations of Berlin and Xi'an. While appearing as an act of communication, offering a garden of new possible ideals, it also physically proves that the Earth is round.

这个穿越地球的大洞把两个位于相反位置的地点——柏林和西安的人们联系到一起，不仅是一种交流，而且打造出新型的花园，同时也证明了一点：地球是圆的。

Ehrenbreitstein Fortress

埃伦布赖特施泰因要塞

LOCATION：Ehrenbreitstein，Germany
项目地点：德国 埃伦布赖特施泰因

AREA：95,000 m^2
面积：95 000 平方米

COMPLETION DATE：2010
完成时间：2010 年

DESIGN COMPANY：Topotek 1 Gesellschaft Von Landschaftsarchitekten Mbh
设计公司：Topotek 1 Gesellschaft Von Landschaftsarchitekten Mbh

Ehrenbreitstein Fortress

The impressive stronghold, overlooking the confluence of Rhine and Mosel rivers is an important national monument. In a comprehensive restructuring of the northern approach as a visitor's facility, the historical plateau is developed as a museum park and the spatial qualities of the site are brought into context with the wider surrounding.

要塞位于莱茵河与摩泽尔河的交汇处，是重要的国家历史建筑。在对北面的通道进行面向游客的综合改造过程中，这片历史高地被开发成了一个博物馆公园，而这个地点的空间性质也被融入更大的环境之中。

The general design concept reorganizes the historic fortress plateau in a spatial and dramaturgic way into the new main entrance of the Fortress of Ehrenbreitstein. The car entrance and the parking lot are placed at the edge of the plateau. Through this the big space keeps monumentality and stages the silhouette of the Fortress to the north. On top of the former barock design construction is a wide lawn laid out which makes out the background for the experience of the Fortress ensemble. A new network of path axes orders the plain and connects up with the existing system of pathways.

总的设计理念是对历史要塞所在的高地进行空间和戏剧式的重新组织，使其成为埃伦布赖特施泰因要塞的新主入口。汽车入口和停车场位于高地的边缘。通过这样规划，这片宽敞的空间保持了其纪念性，同时也向北延伸。原来巴洛克风格的建筑顶部是一片宽敞的草地，成为要塞的背景。新的通道体系将场地规划得井然有序，同时也与原来的通道系统连接在一起。

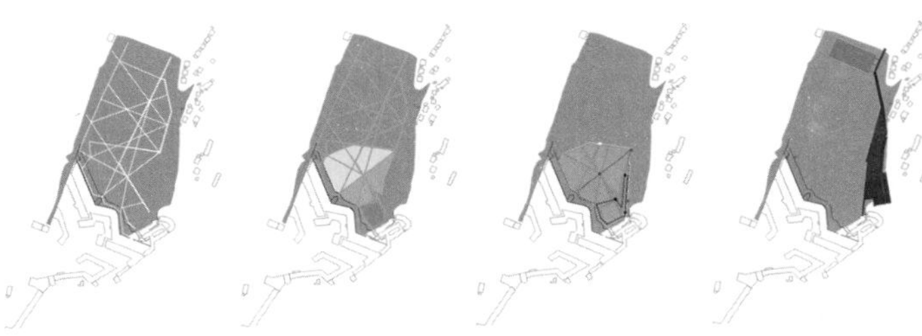

Hackesches Quartier, Berlin

柏林 HACKESCHES 广场

LOCATION：Berlin，Germany
项目地点：德国 柏林

ARAE：5,500 m^2
面积：5 500 平方米

COMPLETION DATE：2010
完成时间：2010 年

DESIGN COMPANY：Topotek 1 Gesellschaft Von Landschaftsarchitekten Mbh
设计公司：Topotek 1 Gesellschaft Von Landschaftsarchitekten Mbh

Hackesches Quartier, Berlin

The urban structure to release the Square is in remembrance to the former church Garnisonskirche.

在城市结构的规划中让出广场的目的是为了纪念以前的 Garnisonskirche 教堂。

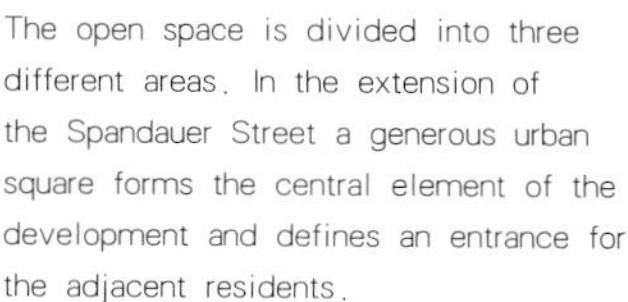

The open space is divided into three different areas. In the extension of the Spandauer Street a generous urban square forms the central element of the development and defines an entrance for the adjacent residents.

空间被分成三个不同的区域。作为 Spandauer 大街的延伸，一个宽敞的城市广场构成本项目的中心元素，同时也成为附近居民的入口。

The character of the surrounding buildings and its central location are emphasized by the high standard of the landscape design. The group of plane trees is surrounded by a bench and defines the setting of the square to the southwest and invites people to stay and to use the square as an amenity place. A prominent group of 3~5 plane trees has been planted in this area.

高标准的景观设计加强了周围建筑的典型特征以及广场的中心位置。几棵法国梧桐被长凳围绕着，成为广场西南方向的"背景"，吸引人们来到广场，为人们带来舒适与便利。广场种植了 3~5 棵法国梧桐，十分显眼。

In between the new Square and the S-Bahn access runs an urban laneway, which is dominated by the local shops without any furniture.

新广场与 S-Bahn 通道之间有一条城市小巷，小巷里主要是一些当地的店铺，几乎没有什么装饰。

The selected material and the colours used in the public open spaces are matching with the surrounding architecture. The material for the square will be white asphalt. The laneway is covered with anthracite-coloured small-sized natural stone paving.

公共空间选用的材料与色彩与周围的建筑相协调。广场将采用白色柏油路面，小巷则用无烟煤颜色的天然小石头铺路。

The whole quarter is framed by a public walkway, which is partly in public and partly in private property and is designed by the typical design guidelines of the city of Berlin.

整个广场由公共人行道围合而成，一部分属于公有，一部分属于私有，按照柏林市典型的设计准则进行设计。

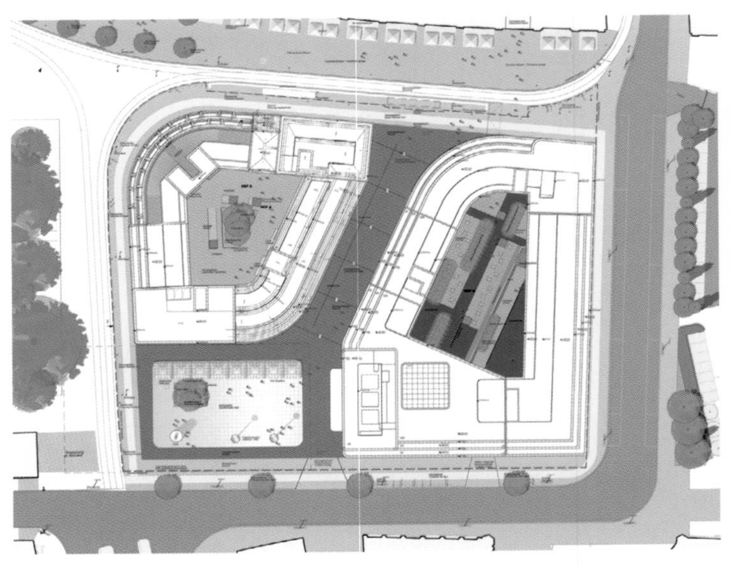

Südkreuz Train Station, Berlin

柏林 Südkreuz 火车站

LOCATION：Berlin，Germany
项目地点：德国 柏林

AREA：50,000 m^2
面积：50 000 平方米

COMPLETION DATE：2006
完成时间：2006 年

DESIGN COMPANY：Topotek 1 Gesellschaft Von Landschaftsarchitekten Mbh
设计公司：Topotek 1 Gesellschaft Von Landschaftsarchitekten Mbh

Südkreuz Train Station, Berlin

The new train station Papestrasse generates an urban internode, which bundles the diverse directions and functions also of the neighbouring quarters.

新建的火车站 Papestrasse 形成了一个城市枢纽，将附近地区的各种结构和功能连合在一起。

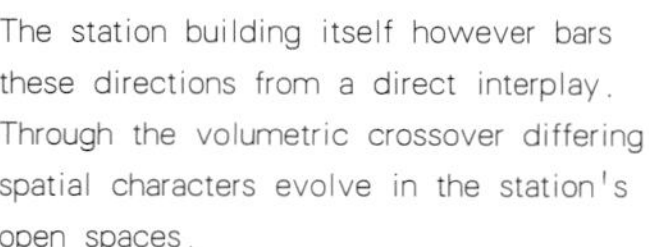

The station building itself however bars these directions from a direct interplay. Through the volumetric crossover differing spatial characters evolve in the station's open spaces.

而火车站建筑本身阻碍了这些结构间直接相互作用。立交桥为火车站的公共空间赋予了不同的空间特性。

Their design assimilates to the general atmosphere of the site without neglecting the specific necessities and aspects of the single space.

设计融入了场地的总体环境，同时又兼顾了每个空间的特定需求及其他方面。

With different designs the four squares are to set spatial accents which reverberate the adjacent city quarters. The interlinkage with these quarters thus becomes the driving theme in the formulation of open spaces at Südkreuz trainstation.

四个广场的设计各具特色，展现出不同的空间特征，与附近的城区相呼应。与这些地区之间的互动于是便成为打造 Südkreuz 火车站公共空间的主题。

The western square, as the main entry towards the district of Schneberg, features a generous openness and a representative character, while at the same time synoptically encompassing the interchange with street-borne public transportation.

作为 Schneberg 地区的主入口，西面的广场十分开敞，而且具有代表性的特征，同时与街道上的公共运输形成互通式的立体交叉。

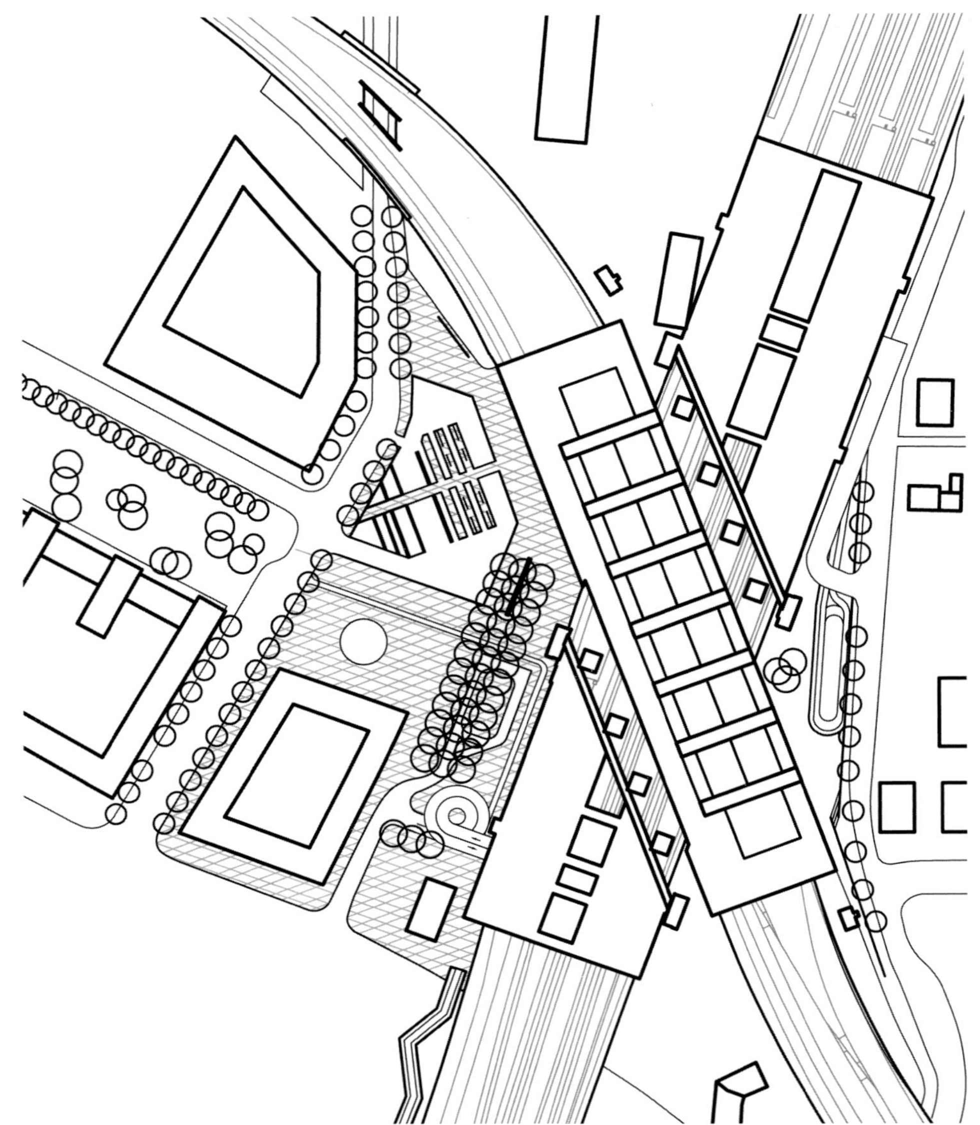

The eastern square is of a more intimate, yet still urban nature. Making use of the topographical situation, a retaining wall parallel to the neighbouring street splits the adjacent pavement level and allows a flowing access to this side of the station. The square emerges as a broadened pavement.

东面的广场显得更加亲切一些，但是仍然具有明显的城市特征。广场利用地形条件，与相邻街道平行的一面挡土墙将相邻的人行道分成不同的高度，提供了一条通往车站这一侧的通道。广场以加宽的人行道的方式出现在人们的面前。

The smaller areas to the north and south serve as secondary entrances to the station. Accordingly, they offer access through a park-like setting with a wound asphalt path in a setting of lawn and loosely strewn trees, which extends vegetation and ambiance.

南面和北面较小的区域是车站的次要入口，提供了通往车站的通道。这里的环境与公园相似，一片草坪和错落有致的树木中有一条蜿蜒的沥青小道，扩大了植被的范围，增强了公园的氛围。

ASPECT Studios

Chris Razzell (Executive Director)

Bachelor of Landscape Architecture,
Royal Melbourne Institute of Technology
Registered Landscape Architect,
Australian Institute of Landscape Architects

Chris is the founder and Director of ASPECT Studios. Chris' s business philosophy is founded on the principle of studio individuality supported by a national and international network of expertise. Chris is driven by a commitment to delivering the highest quality design
outcomes across the breadth of landscape architecture and urban design. His inventive, disciplined skills and integrated involvement are brought to bear on anything from team management, the design business, fnancial, logistical or strategic issues, to fostering the potential of all ASPECT Studios staff.

ASPECT Studios

We do landscape architecture, urban design, interactive digital media. We do excellence, difference, passion, confdence, ambition and bravery.
We work with government, the private sector and creative professionals. We're designers, we're collaborators and we do innovation. We win awards, not just for what we design but for the way we think.
Sustainability: it's central, it's critical, and we aim high, seeking the best outcomes for your project. We look at the project life-cycle in economic, social, environmental and cultural terms.
We are studios: That means one group working in teams, collaborating, debating, learning, fostering and growing, creating opportunity. Through this we produce considered project-specifc solutions for you.
We work with the best, in the arts, the sciences, the technical professions. That's best practice. We adapt. We respond.
We're defned and built by our core people who are experienced, fexible and diverse. We are Australian owned. We have local knowledge, ensuring we can deliver the most innovative and cost effective solutions to you.

8a The Terrace, Birchgrove

Birchgrove 8a The Terrace 庭院

LOCATION：Sydney, Australia
项目地点：澳大利亚 悉尼

PHOTOGRAPHER：Simon Wood
摄影师：Simon Wood

DESIGN COMPANY：ASPECT Studios
设计公司：澳派（澳大利亚）景观规划设计工作室

8a The Terrace, Birchgrove

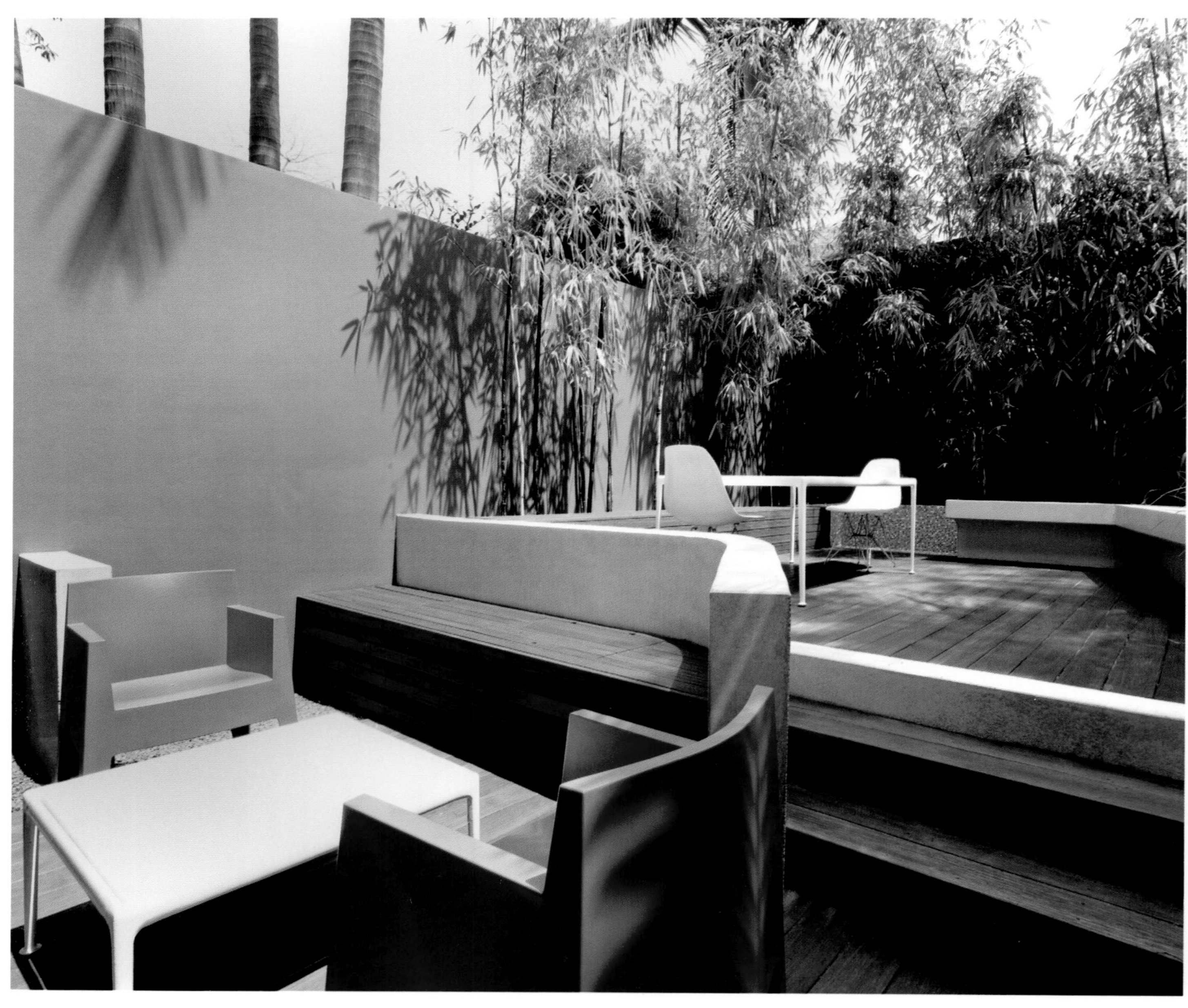

An intimate courtyard creates an oasis in an inner city neighborhood.

一个私家庭院营造了一片城市内部的绿洲。

Giant black bamboo, one-off in-situ concrete details and Australian native hardwood create a beautifully considered external space. Multi-use spaces are created by the layout of these wel-crafted elements. Sculptural concrete walls double as seating back rests and mediate level changes. Timber benches conceal storage areas. A cantilevered concrete bench to the rear of the garden frames the outdoor shower and becomes a striking feature when lit at night.

高大的紫竹、优雅的弧线景墙、澳大利亚本地硬木铺装，打造出一个美丽的外部空间。通过这些景观元素的巧妙设置，使得这个有限的空间营造出多个多功能室外活动空间。弧线混凝土挡土墙不仅造型精美，而且同时又有很好的功能性，既是木平台座椅靠背，又巧妙地抬高了局部的地形，丰富了景观层次感。此外，设计师还在空间的利用上动脑筋，如在木座椅下方做一个储物收纳空间；利用一道混凝土挡墙划分了户外淋浴的空间，再通过射灯的投影使这面墙形成一个视觉焦点。

Timber uprights and timber arbour help to integrate the internal and external living spaces and provide a seamless space. The high-quality material and planting palette make the garden a highly desirable space to live in.

木质廊架与射灯结合，巧妙地让室内空间与室外空间有一个自然的过渡，无论白天还是夜晚景观效果都那么精美融合。庭院景观选用高品质的材料，再与精美的植被相结合，打造出一个宜居、舒适而又精美的私家庭院。

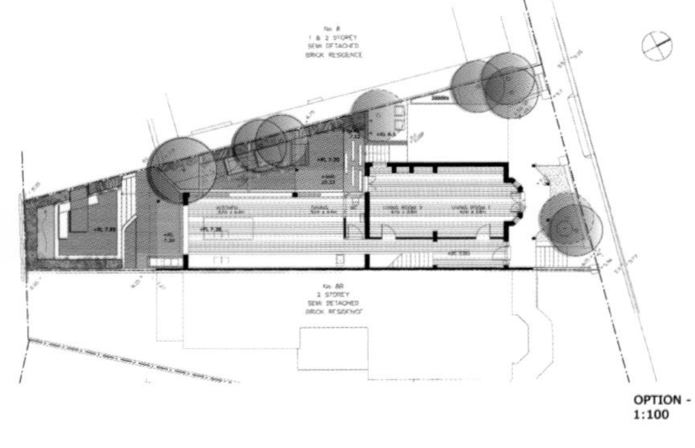

717 Bourke Street

717 号 Bourke 大街景观设计

LOCATION：Melbourne，Australia
项目地点：澳大利亚 墨尔本

AREA：10,000 m^2
面积：10 000 平方米

COMPLETION DATE：2010
完成时间：2010 年

TEAM：Metier3 Architects
设计团队：Metier3 Architects 建筑设计公司

DESIGN COMPANY：ASPECT Studios
设计公司：澳派（澳大利亚）景观规划设计工作室

717 Bourke Street

ASPECT Studios was engaged to design and document the streetscape, pocket park, podium forecourt and courtyards for this mixed use development in the Docklands.

澳派景观设计工作室受业主邀请，对这处位于墨尔本港口区的商业综合体项目提供景观设计，包括街道景观、迷你公园、屋顶平台和庭院的设计。

The landscape designers work collaboratively with Metier3 Architects and PDS Group, and a new precinct has been added to the Docklands.

景观设计师同建筑设计公司Metier3 Architects 以及业主PDS 集团一同开展工作，成功地打造一个独特的商业景观空间，成为墨尔本港口区又一个新地标。

Wanting to remove the typical 90 degree relationship with landscape and the building, ASPECT created a language of tectonic landscape forms, made of ramps, seats, garden beds and decks.

设计避免了景观与建筑常规的90度直角的空间关系，创造出一种独特的景观语言，成功地打造了一系列商业景观功能空间，包括梯形的景观地形、座椅、种植池和平台等。

Foley Park

Foley 公园

LOCATION：Sydney，Australia
项目地点：澳大利亚 悉尼

AREA：6,000 m^2
占地面积：6 000 平方米

PHOTOGRAPHER：Florian Groehn
摄影师：Florian Groehn

TEAM：ASPECT Studios (Lead Consultants)，CAB Consulting，Fiona Robbé and TTW
设计团队：ASPECT Studios（主创设计师），CAB Consulting，Fiona Robbé and TTW

COMPLETION DATE：2009
完成时间：2009 年

DESIGN COMPANY：ASPECT Studios
设计公司：澳派（澳大利亚）景观规划设计工作室

Foley Park

ASPECT Studios was commissioned by the City of Sydney to undertake design development and subsequent documentation for the upgrade of Foley Park. The result is a carefully designed contemporary intervention within this existing heritage park in the heart of Glebe which provides a revitalised community space for local residents.

ASPECT Studios澳派景观规划设计工作室受悉尼市议会的邀请，对悉尼Glebe区的Foley公园进行景观改造。设计师经过仔细推敲，采用现代的设计语言，为当地居民提供了一处生机勃勃的户外空间。

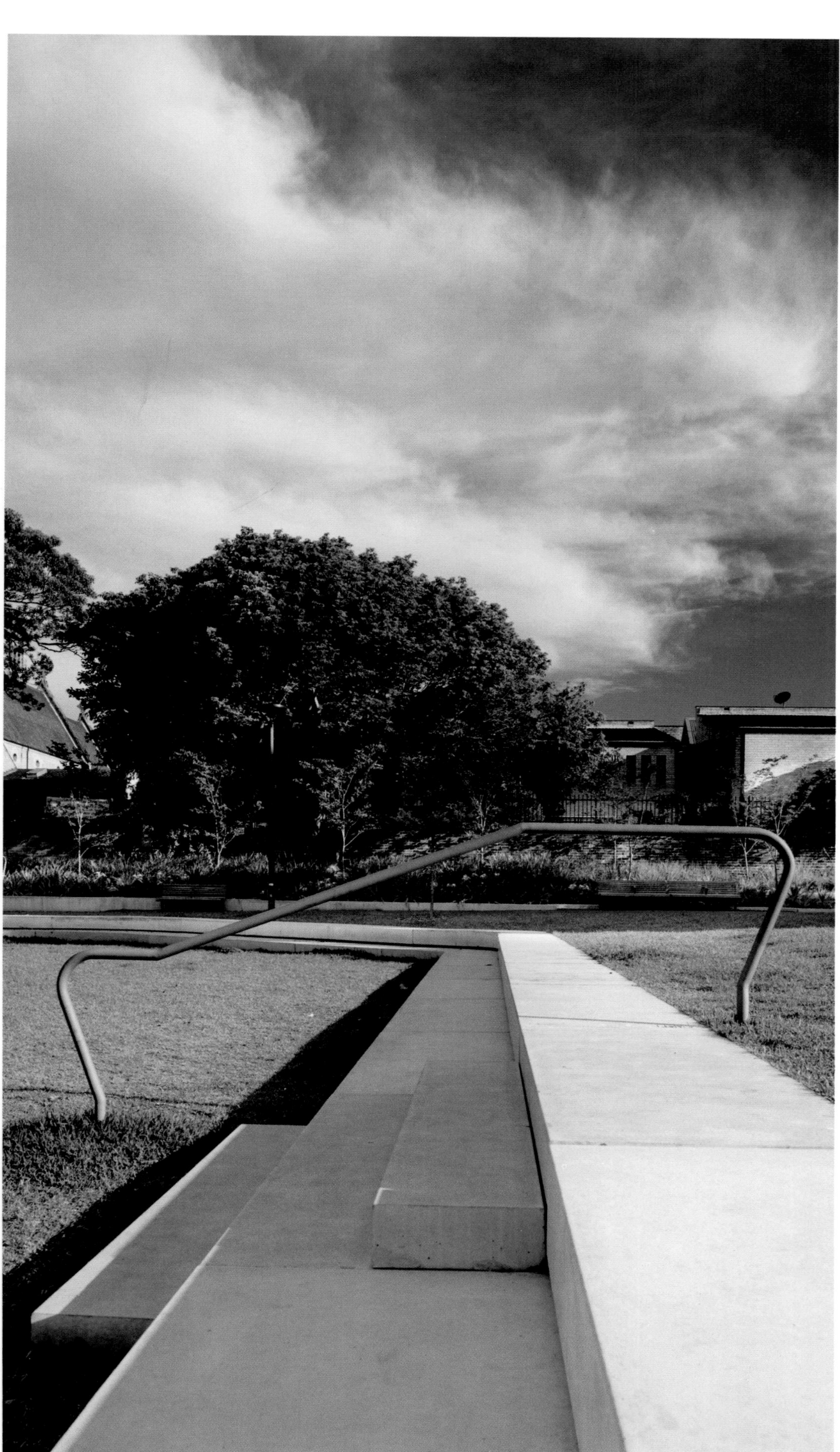

Foley Park is a 6,000 m^2 urban park located at the intersection of Glebe Point Road and Pyrmont Bridge Road in Sydney. It is the largest local park in Glebe. Foley Park offers a complex and constrained design challenge. In light of this, the design response provides maximum community space and park amenity whilst being highly sensitive to the existing park elements.

Foley公园占地6 000平方米，位于悉尼Pyrmont桥路和Glebe路的交叉口，是Glebe区最大的公园。鉴于公园复杂的历史和人文因素，设计中存在许多限制和难点。在谨慎考虑公园内现有景观元素的同时，设计最大程度地打造了户外聚会空间，营造出舒适且宜人的静谧氛围。

The design builds on the Plan of Management prepared by the City of Sydney and utilises extensive site analysis, historical analysis and community consultation.

根据业主提供的改造目标，设计师进行了大量的场地分析、文化分析，并且咨询了社区居民的意见。

Design upgrades include improvements to all street frontages with widened entries on Glebe Point Road and Pyrmont Bridge Road. Existing pathways are reconfigured, materials and planting improved. The new park layout creates a series of linked zones: the "Village Green" the "Hereford House area" and the "Play Precinct".

景观改造包括了所有街道临街面的设计，同时拓宽了Glebe Point路和Pyrmont桥路的入口。步行道也进行了重新设计，改进了铺装材质和种植配置。改造后的公园打造了3处相互串联的空间，包括“绿色村庄”，“Hereford House区域”和“游戏区”。

The Village Green is a large turf area which is regraded and simplified to provide a dramatically improved park structure and increased amenity. This area works with the existing "sheltered oasis" quality of the main lawn area enclosed by large fig trees. The existing turf is made flatter, enlarged, cleared of small trees and shrubs and defined by wide pre-cast concrete edges that function as sitting steps on the southwestern side. This turf area allows for greater passive recreation uses such as casual ball games and community events to be accommodated.

"绿色村庄"是一大片宽阔的草坪区，设计师重新调整了草坪的坡度，展示出一处更清晰的空间布局。设计平缓了原有的坡度，增大了草坪区域的面积，移除了场地内的乔木和灌木，在草坪西南面新增了宽阔的预制混凝土座墙。改造后的区域形成了一处亲密休憩环境。

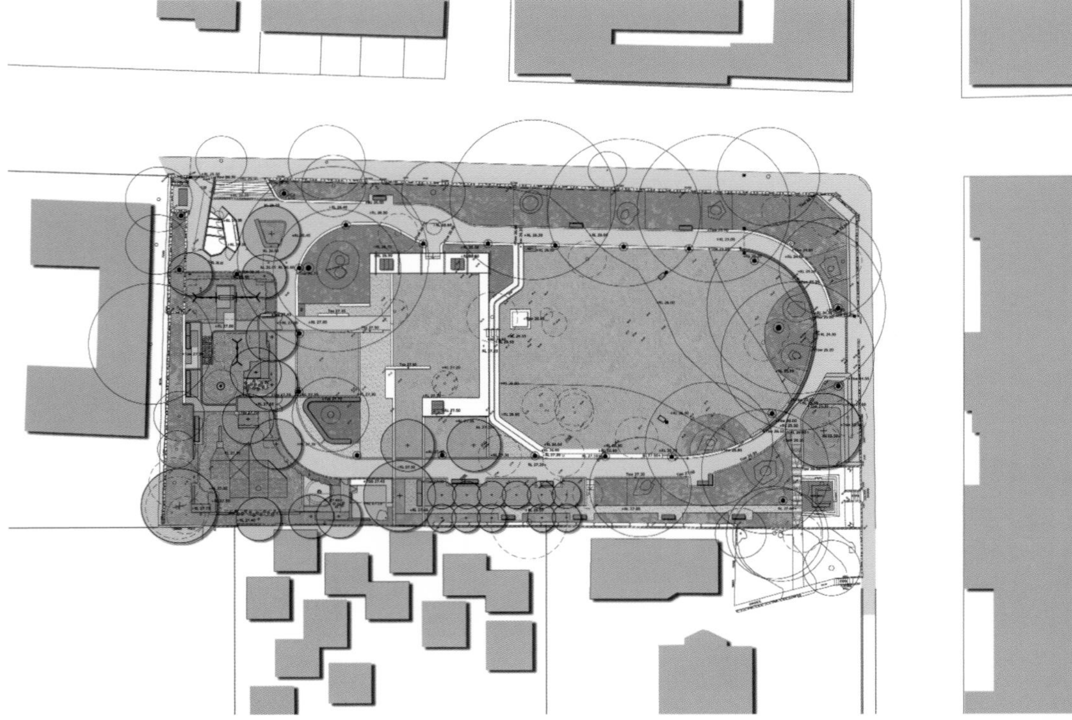

The Hereford House precinct is designed to provide a transition between the play area and the less-structured Village Green and to allow for multiple uses. It exploits the topography of the site and reinstates the former house position within the greater park area as the central focus of the park.

Hereford House 区域是游乐区和绿色村庄之间的过渡空间。设计充分利用了现场原有的地形，并将一处历史遗迹建筑形成了整个公园的重点。

The extensive new playground and interpretive elements create a meaningful local focus. The character of the play area is defined by the coloured poles of the play equipment inter-dispersed with low planting and clean-trunked trees. The rubber softfall ground plane has been designed as a dappled green and grey carpet with the equipment located in fine mulch.

宽敞的游乐场和精致的景观元素形成了一处别有韵味的视觉焦点。游乐区域内色彩丰富的立柱散置在周边低矮的灌木和高杆乔木之间。绿色和灰色系的橡胶铺地宛若地毯一般铺设在游乐区域内，游乐器材放置在木屑铺地内。

40

Melbourne Convention Exhibition Centre

墨尔本国际会议会展中心

LOCATION：Melbourne，Australia
项目地点：澳大利亚 墨尔本

AREA：10,000 m^2
面积：10 000 平方米

COMPLETION DATE：2009
完成时间：2009 年

DESIGN COMPANY：ASPECT Studios
设计公司：澳派（澳大利亚）景观规划设计工作室

Melbourne Convention Exhibition Centre

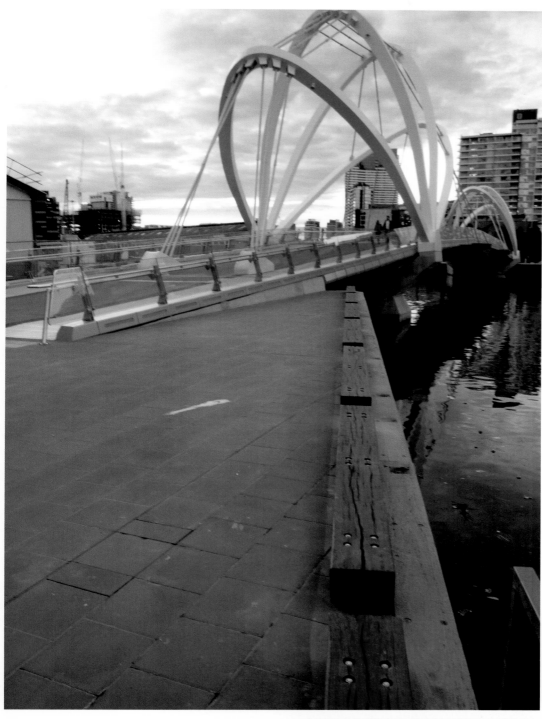

ASPECT Studios was commissioned by the Plenary Group, Multiplex and Contexx to undertake the design and documentation of the new public realm for the Melbourne Convention Centre and South Wharf Precinct. The Melbourne Convention Centre is the first convention centre in the world to achieve a 6 Star Green Star environmental rating.

应 Plenary 集团、Multiplex 公司及 Contexx 公司的邀请，澳派（澳大利亚）景观规划设计工作室为墨尔本会议中心及南码头公共区域提供景观设计。墨尔本会展中心是全球第一个六星级的绿色建筑。

The landscape design for public realm places include: Convention Centre; Hotel; Retail; Residential; Public forecourts; Promenades and laneways; Refurbishment of the existing Exhibition Centre Park; Heritage; landscapes.

公共区域景观设计的范围包括：
会展中心、酒店、零售区、住宅区、公共广场、滨水人行道、会展中心公园的改造与重建、历史景观的利用与重建。

Key principles of the Public Realm: 24—hour public access throughout the precinct; Clear and direct pedestrian and cyclist routes; An attractive Yarra River; Promenade; Interesting spaces with the precinct for residents, pedestrians, visitors, shoppers and delegates; Connection of the precincts areas to the Yarra River; Incorporate of all heritage elements.

公共空间景观设计的主要原则有：
保证地块内全天候的便捷通行；设立清晰的人行道和自行车道系统；打造雅拉河畔滨河景观；打造散步长廊为居民、游客、购物者、参观代表等不同的人群提供不同的功能设施；处理会展中心至雅拉河的连接与通道；将包含所有的遗产元素。

Public Realm Surfaces:
The primary pavement surface is "bluestone" pavers, with secondary pavements of reconstituted granite pavers and "bluestone" matching high-quality concrete pavers. Other surface materials will be used in smaller and periphery areas such as fine gravel, granite sets, permeable paving surrounding trees, coloured concrete and asphalt. Throughout the commercial development tactile pavements will be used as required to mark hazards, level changes, and pedestrian crossing points.

公共区域铺地：
路面主要的铺装材质是青石，其次是再生花岗岩铺地，以及与青石相配的高档混凝土铺地。其他小型区域和周边地区采用不同的铺地材料，如小鹅卵石，花岗岩格，乔木周围可渗透铺地，彩色混凝土铺地和沥青等。整个商业区使用的可触知的铺地，标出危险区域、上下台阶和过街人行道，方便盲人的使用。

Vegetation principles:
Continuity of species along Yarra Promenade;Low water—use plants;High canopy species for shade;Summer shade and winter sun in confined spaces.

植被选用原则：
延续雅拉河林荫道使用的植物类型，采用水量吸收少的植物；大树冠且可遮荫的植物、夏日遮阴，冬日透光的植物。

ASPECT Studios is proud of making a significant contribution to the environment in the landscape design in making sure that the stormwater on site is collected to provide treated water for toilet flushing, irrigation and cooling towers.

澳派工作室为在本项目中对环境作出的努力感到自豪。在会展中心的景观设计中积极融入了环保的做法，收集场地内的雨水，处理后用于卫生间用水、植被的灌溉和建筑的降温。

Shibagaki and Ng Residence

Shibagaki & Ng 别墅花园

LOCATION：Sydney，Australia
项目地点：澳大利亚 悉尼

AREA：1，000 m^2
面积：1 000 平方米

COMPLETION DATE：2009
完成时间：2009 年

TEAM：ASPECT Studios，Marsh Cashman Koolloos（MCK）Architects
设计团队：澳派（澳大利亚）景观规划设计工作室，Marsh Cashman Koolloos 建筑设计工作室

Shibagaki and Ng Residence

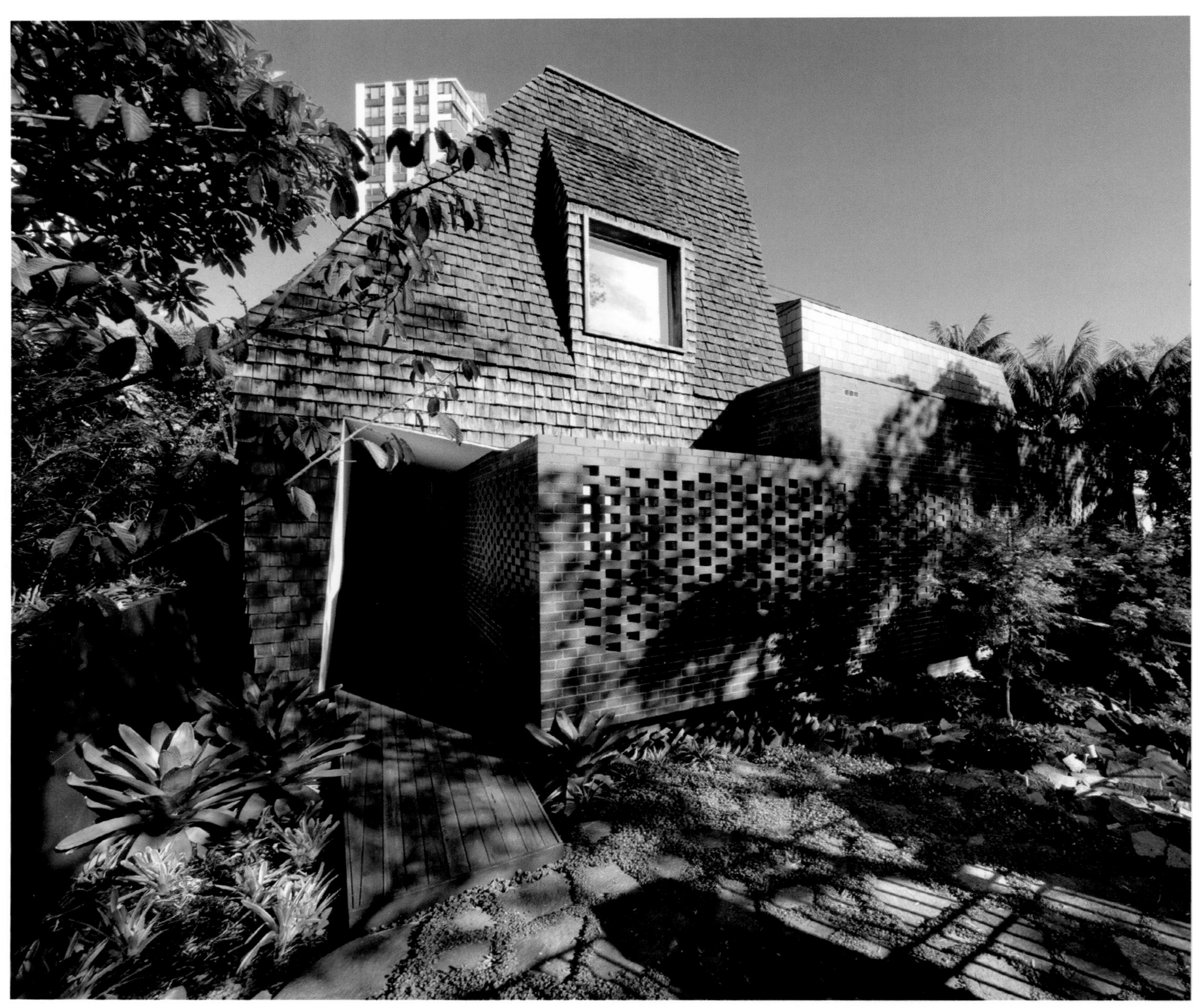

A unique fusion between architecture and landscape is here the design intent that is to blur the line between internal and external space. Floor surfaces extend from inside the home out into the garden spaces. All species are drought–tolerant and hardy once established. Water is recycled from the roof and stored in a concealed tank at the front of the property. Landscape materials are "true materials", a robust and simple palette, durable and minimised the need for complex maintenance regimes.

别墅花园设计的成功之处在于设计师巧妙地将室外景观与室内景观进行自然的过渡与融合。室内的铺地一直延续到户外的花园空间。庭院花园选用耐寒、低维护的植物。屋顶的雨水全部都收集起来，并储存在别墅前方隐蔽的雨水箱中。景观采用选用有质地感的石材，简洁、耐久，便于日后的维护。

Sydney Olympic Park Jacaranda Square Landscape Design

悉尼奥林匹克公园 Jacaranda 广场景观设计

LOCATION：Sydney，Australia
项目地点：澳大利亚 悉尼

AREA：5,000 m^2
面积：5 000 平方米

COMPLETION DATE：2008
完成时间：2008 年

PHOTOGRAPHER：Simon Wood，Sacha Coles
摄影师：Simon Wood，Sacha Coles

Sydney Olympic Park Jacaranda Square Landscape Design

TEAM：ASPECT Studios Pty Ltd，MWA，Deuce Design
设计团队：澳派，MWA，Deuce Design

DESIGN COMPANY：ASPECT Studios
设计公司：澳派（澳大利亚）景观规划设计工作室

Decorative coloured brick walls with concrete lounges, colourful canopies, recycled brick ground plane and native trees combine with a large green platform to create Sydney Olympic Park's memorable new community space.

装饰性彩色砖墙，混凝土材质的长椅，色彩缤纷的雨篷，循环再造砖块的铺地，本土的树种与一个大型的绿地空间，构成了悉尼奥林匹克公园令人难忘的新社区空间。

Jacaranda Square, "The Everyday Stadium" is the first new public space of Sydney Olympic Park to sit within the framework of the 2025 vision, to build on the Olympic legacy and to create a vibrant, active and sustainable town centre. The project is a result of a design competition won by a group of like-minded professionals in the fields of landscape architecture, architecture and graphic design.

Jacaranda 广场，“天天运动场”是悉尼奥林匹克公园第一个新的公共空间，属于悉尼 2025 年远景规划的一部分，在奥林匹克遗址内建成一个生动积极的环境可持续中心。澳派在此项目荣获设计竞赛第一名，项目由一个由志同道合的景观设计师，建筑师和平面设计师组成的设计团队共同完成。

In 2004, ASPECT Studios invited McGregor Westlake Architects and Deuce Design to collaborate on a limited design competition to design the new open space at the heart of the Sydney Olympic Park town centre.

2004 年，澳派工作室与 McGregor Westlake 建筑事务所和 Deuce 设计公司共同合作，赢得了悉尼奥林匹克公园中心的新公共空间有限设计竞赛第一名。

The resulting award-winning scheme is a new urban park for passive recreation and community gathering. The term "The Everyday Stadium", is both a gentle, ironic nod to the Olympic legacy and a description of the design concept, which is made up of 3 elements; a large central open space, an edge of walls and seats, and large perimeters of shade—one built and the other through trees.

获奖项目的主题是设计一个供人们静态休闲和市民聚会的新都市公园。公园的主题“天天运动场”就意味着这个新绿地是奥林匹克精神的传承，同时也是设计概念的表述。共由 3 个元素组成：一个大型中央开放空间，围墙和座椅以及大型的遮荫设施——由一个建成的遮阳篷和高大树冠的乔木组成。

Brick was used extensively in the project, in part due to the site's proximity to the former Homebush brick pit and to give the park a dynamic, textural and colourful character.
Glazed bricks were used as a cladding on the perimeter walls, interspersed with Austral Gertrudis to give a striking visual pattern.

项目设计广泛使用砖块，部分原因是场地接近霍姆布什砖窖的旧址，而且砖块的运用能给公园产生一种动态的、充满质感和色彩缤纷的效果。
广场外围围墙使用彩色釉面砖，与澳洲Gertrudis地区的风格互相融合，展现一种鲜明的视觉效果。

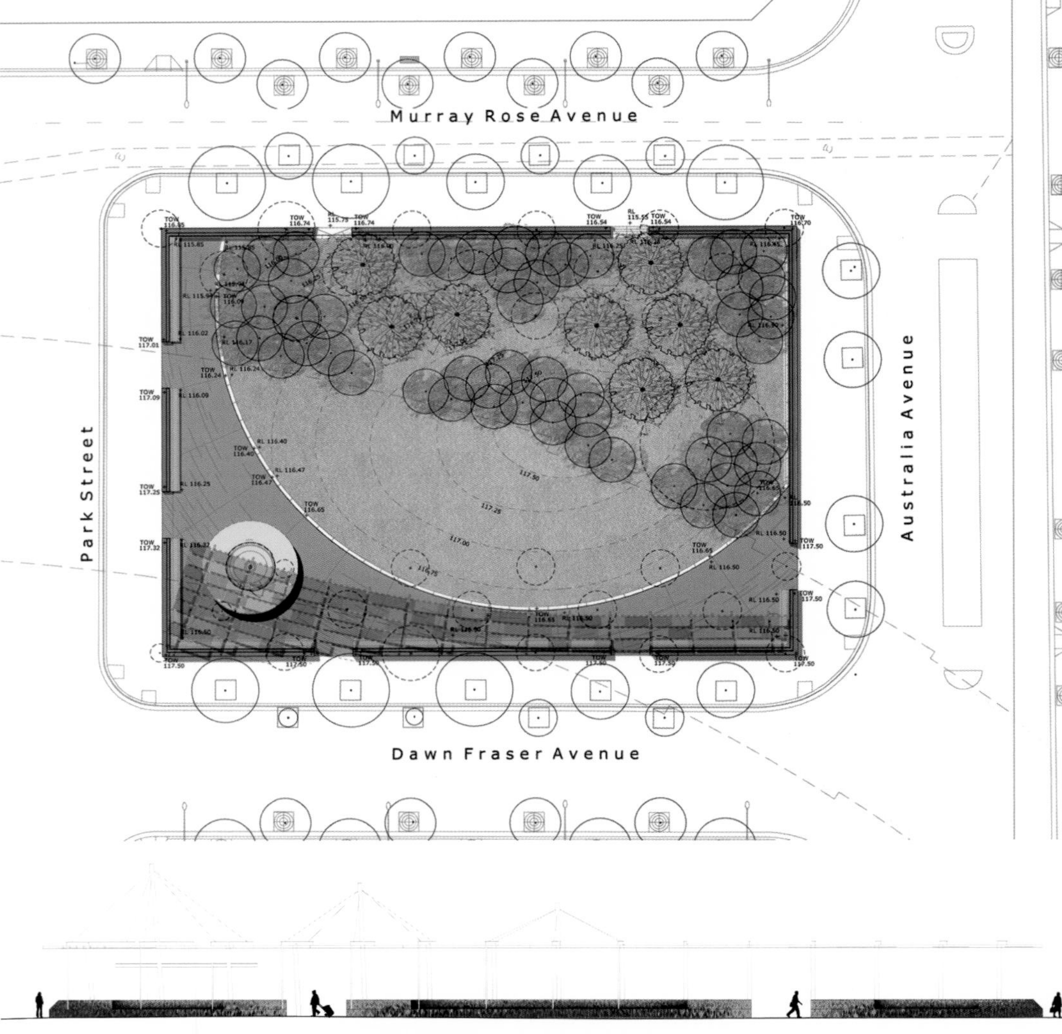

This effect was coined the "Brixel", which is a pixilated pattern realised through the use of 4 different coloured bricks. Bowral Blue Bricks clad ded the stairs and ramps, and Bowral 50's were used as paving highlights to accentuate both the concrete set-out and to create rays stemming from the origin point.

通过4种不同颜色的砖块，形成一个点状的特色图案。在楼梯和斜坡使用南威尔士洲独有的Bowral 50's系列砖块，形成铺地的亮点，既强调了混凝土的排列也营造出从圆点发射出光线的视觉效果。

Recycled bricks were used on edge in a concrete stretcher pattern for the paved area adjacent to the cafè. The recycled bricks harmonise their rustic quality with the sharply-crafted adjacent precast-concrete elements.

咖啡厅附近区域的硬铺区域使用环保砖作为修饰用途。环保砖的质感可以软化周围预制混凝土给人带给的生硬的感觉。

The final result is colourful, clean and green visual effect. The design features a series of modular precast–concrete lounge suites; a canopy of polychrome greens; walls of glazed pixilated bricks set amongst a landscape of native trees.

广场最终呈现出色彩缤纷、清洁和绿色的视觉效果。设计包括一系列规格化的预制混凝土休息套房；颜色丰富的绿色遮阳篷；特色彩色釉面砖铺面的围墙，本地的树种，构成了广场独特的景观效果。

Green ecolgical theory is in the design, and the park features include: a recycled brick pavement, recycled materials and recycled water for irrigation combined to achieve a meaningful, environmental and socially sustainable place.

在广场的设计中综合了一系列的绿色生态的理论：一个环保砖铺装路面，使用循环再造的材料，收集雨水用于灌溉，从而从真正意义上的做到环境和社会可持续发展。

This is a "complete project", which successfully fuses landscape architecture, Industrial and graphic design with architecture to create an intelligent and memorable open space.

这是一个“完美的项目”，成功显示景观设计、工业及平面设计与建筑设计的完美结合，是一个充满智慧、令人难忘的开放空间。

Sydney Rhodes Lot 8 Tandara – Little Space, Big Courtyard

悉尼 Rhodes 8 号地块 Tandara 庭院——小空间、大花园

LOCATION：Sydney，Australia
项目地点：澳大利亚 悉尼

AREA：500 m²
面积：500 平方米

COMPLETION DATE：2007
完成时间：2007 年

CONSULTANT：Mirvac Design
顾问：Mirvac Design

DESIGN COMPANY：ASPECT Studios Pty Ltd
设计公司：澳派（澳大利亚）景观规划设计公司

TEAM：ASPECT Studios Pty Ltd，MWA，Deuce Design
设计团队：澳派、MWA，Deuce Design

Sydney Rhodes Lot 8 Tandara – Little Space, Big Courtyard

A carefully-selected planting palette and an intricate series of paths create a lush, richly-textured communal courtyard for a residential development in Rhodes.

精心考虑的植物配置、不同层次的园林小径交错融合，为 Rhodes 的住在小区营造出一个郁郁葱葱，富有质感的庭院花园。

The landscape design for the communal courtyard at Rhodes Lot 8 Tandara works with an awkward site geometry to create a seamless series of courtyard spaces. The scheme incorporates a range of conditions including turf areas, richly-textured planting zones and groves of trees.

Rhodes 8 号地块的 Tandara 庭院的景观设计打破了凌乱的地块的限制性，巧妙地将现场地块变废为宝，对草坪、花槽、树阵等景观元素进行巧妙的布局，打造出一个流畅而又富有层次感的庭院空间。

A series of interconnected pathways link private courtyards to entrance lobbies, and create an intricate overlay of small communal spaces.

花园小径从入口大堂交错延伸，围合出一个个庭院的小型活动空间。

Two tree species are used across site in groves and clusters. The species were carefully selected to provide a strong year-round visual aesthetic and scale, as well as a visual screen for surrounding apartments. In the corners of the courtyard, Smooth-leaved Quandongs (Elaeocarpus eumundii) are planted in dense groups whilst Rough Tree Ferns (Cyathea australis) are located in the heart of the space. These tree ferns with their distinctive foliage and colour provide a beautiful pattern of shadow and light to the communal space.

庭院的植物配置选用两种树形优美的骨干乔木，分丛而种，保证全年景观上的视觉美感，同时也保护了周围住户的私密性。在庭院的角落处，密集种植树叶油亮光滑的杜英，而庭院的中心处种植澳大利亚本土的蕨树。植物的叶子和颜色形成了一个美丽的庭院花园光影空间。

Materials are carefully selected to enhance the distinctive feel of the courtyard. Permeable paving allows for water infiltration into deep soil areas. Custom timber furniture is integrated into planter walls, providing areas for rest and respite.

庭院景观材料的选用也经过仔细考虑，突出庭院的独特风格。透水砖的选用能保证雨水可以自然渗透到深土区。庭院的小品均为设计定做的木质小品，让挡土墙、种植框与座椅一体化，保证空间的使用率，为住户提供一个休息观景的好去处。

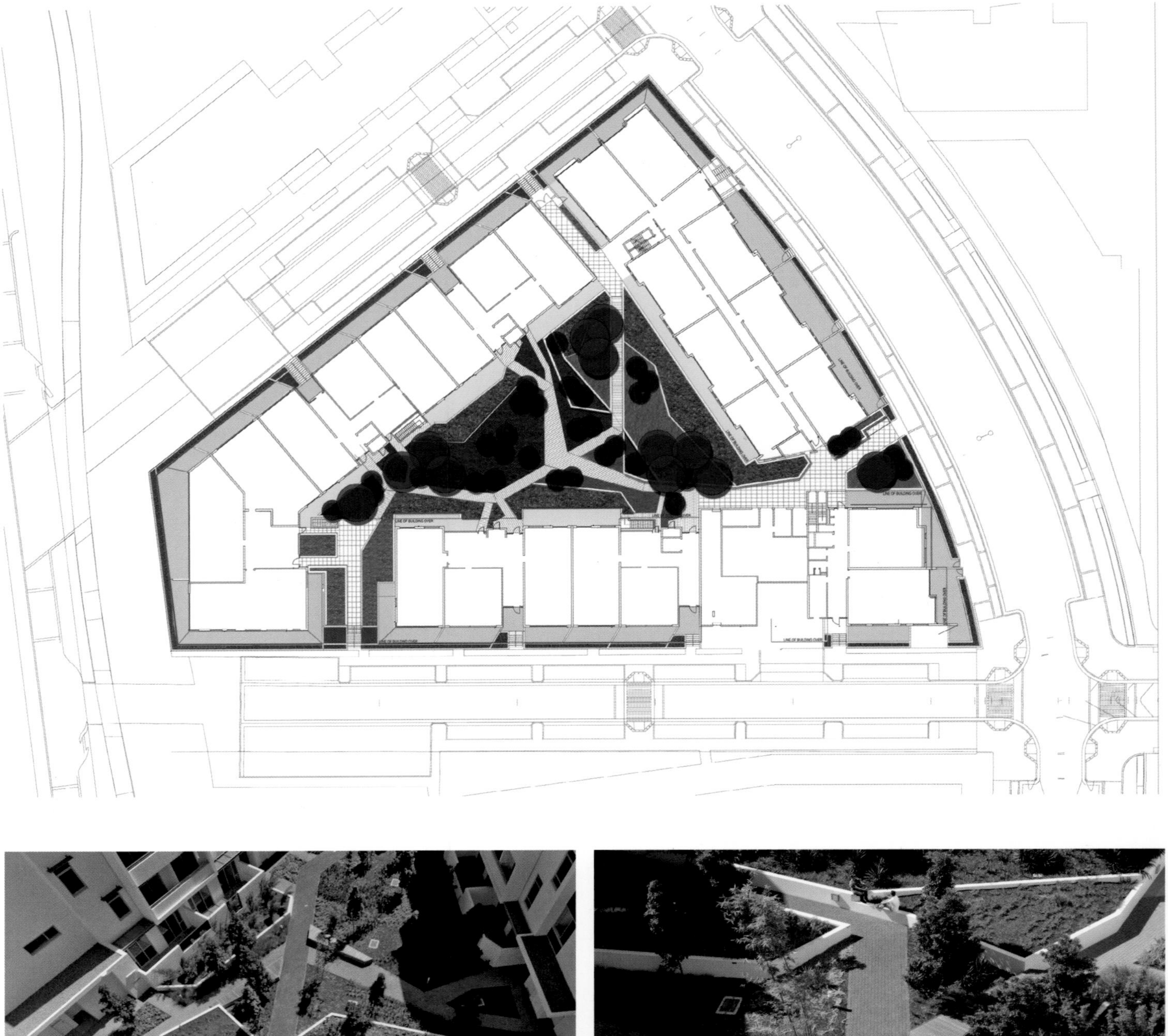

The National Emergency Services Memorial

国家紧急救灾服务纪念馆

LOCATION：Canberra，Australia
项目地点 ：堪培拉 澳大利亚

AREA：10,000 m^2
面积 ：10 000 m^2

AWARD ：Victoria Aila Awards 2004
奖项：2004 年度澳大利亚景观设计师协会维多利亚州优秀设计奖

DESIGN COMPANY：ASPECT Studios Pty Ltd
设计公司：澳派（澳大利亚）景观规划设计公司

The National Emergency Services Memorial

ASPECT Melbourne in 2003 won the National Design Competition for a new civilian memorial to the Emergency Services and Personnel of Australia. The Memorial is one of the first civilian memorials to grace the shores of Lake Burley Griffin and compliments the traditional military memorial axis of Anzac Parade.

澳派墨尔本分公司于2003 年在澳大利亚堪培拉国家紧急救灾服务纪念馆国家设计竞赛中顺利中标，项目的发展目标是为了纪念澳大利亚为国家光荣牺牲的抢险救灾人员。本项目是澳大利亚第一个民用纪念馆，坐落在Burley Griffin 湖畔，与澳大利亚军人战争纪念馆相得益彰。

It draws on the outdoor experience of many cataclysmic events, through perhaps subliminal recollection of such commonly-shared and emotive Australian moments as lines of grass fire at night, lightning flashes, the shadows of one's body cast by strong sunlight or fire.

设计的灵感源于人们对于户外灾难的经历，来源于潜意识中对事件的回忆，如晚上着火的草地、闪电划过的夜空、被强光或火光投射的人影等。

The shards of red material embedded into the slope of the hill, the strong shadows cast by sun and fire, the drawing of the eye across the wider landscape, are all experiences felt by the Service workers at moments during and after a crisis. Ultimately, however, the direction into and onto the "Blanket" brings the visitor to experience the sense of safety provided by the Emergency Services, in the enclosure of the memorial space itself.

斜坡上镶嵌红色的碎片，火焰与太阳拖出人们长长的影子，被更广阔的景象吸引的眼睛这些细部设计能让参观者感受到危机发生时和发生之后投入救援服务的工作人员的思想状态。然而，在参观完浮雕墙后，纪念馆的围合空间又给参观者带来安全感。

The spirit of the emergency services is embodied in this single, powerful gesture of the memorial.

国家紧急救灾服务纪念馆通过简洁而又令人震撼的设计，体现抢险救灾者的奉献精神。

In one instance the memorial aggressively breaks from the gentle continuity of the surrounding landscape, evoking the catastrophic events that call the Emergency Services into action. And on the other, the memorial becomes an inward-folding form offering protection and comfort to all Australians during these times of tragedy.

一方面，纪念馆打破了周围景观的连续性，暗示着灾难的发展，急需国家紧急救灾服务立即行动。另一方面，纪念馆采用内折的形式，暗示着在灾难来临的时刻，国家紧急救灾服务能为所有澳大利亚人民提供保护和安慰。

Sydney Wetland 5 ESD + Landscape Design

悉尼 5 号湿地生态设计与景观设计

LOCATION：Sydney，Australia
项目地点：澳大利亚 悉尼

AREA：20,000 m^2
面积： 20 000 平方米

COMPLETION DATE：2007
完成时间：2007 年

PHOTOGRAPHER：Simon Wood，Sacha Coles
摄影师：Simon Wood，Sacha Coles

AWARDS：CCAA Commendation Award 2009
AILA NSW Award for Commendation in Design 2007
获得奖项：2009 年度澳大利亚水泥与混凝土运用协会优秀设计奖
2007 年度澳大利亚景观设计协会新南威尔士洲优秀设计奖

LANDSCAPE ARCHITECT：ASPECT Studios
景观设计公司：澳派（澳大利亚）景观规划设计工作室

Sydney Wetland 5 ESD + Landscape Design

ASPECT Studios was commissioned to design and document Wetland 5, the culmination of the chain of wetlands in Sydney Park. This existing area is the oldest parcel of land in Sydney Park and the only piece which remains on deep soil (rather than on fill). The condition of the wetland was undermining the environmental benefits of the wetland system. ASPECT Studios brought extensive site knowledge to the project, having completed the detailed master plan for Sydney Park in May 2006.

澳派受到悉尼市议会的委托，为悉尼公园湿地链 5 号湿地进行景观设计和施工图设计。项目现场是悉尼公园现存的唯一拥有深层土（而非回填土的）的地块，也是悉尼公园最古老的地块。现有的湿地不能全面发挥其对环境的促进作用。澳派在 2006 年 5 月圆满完成悉尼公园的详细总体规划设计后，为本项目带来了丰富的现场知识。

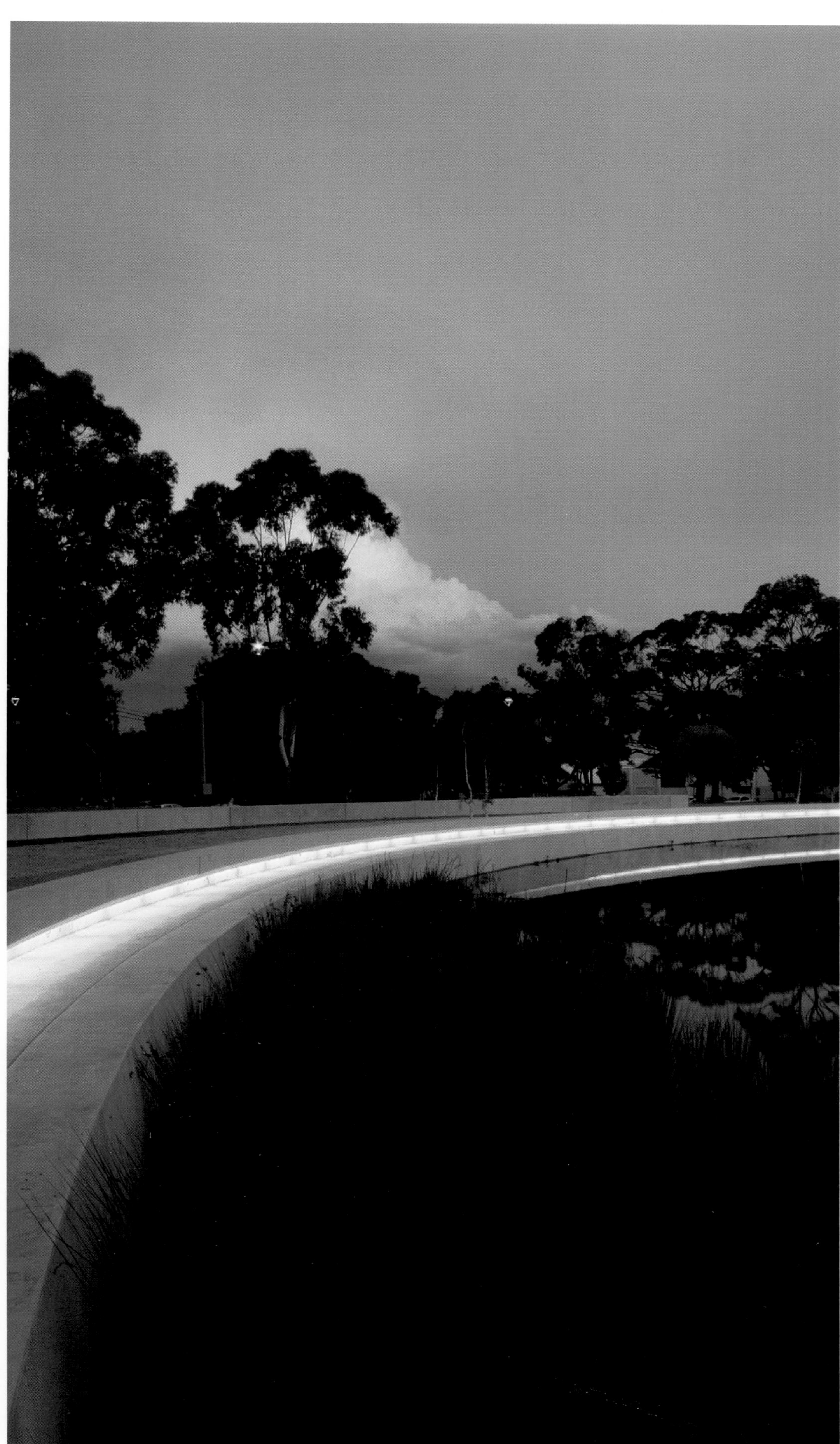

The scope for the project included design development and construction of the wetland and surrounds including pathways, retaining walls, seating and shade trees. The materials selected reflected the vision of the detailed master plan which was to recognise and mark this place as the culmination and holding point for the greater parks wetland system. The design of the high quality, in-situ concrete walls worked as a semi circular bracket or "frame" to the park water course. ASPECT Studios saw the opportunity to mark the upgrade of the park by inserting a simple and robust gesture which is of its time.

项目的设计范围包括湿地及其周边的深化设计和施工图设计，包括道路，挡土墙，座椅和遮阴乔木的设计。所有的材料都是根据总体规划设计的风格来选择，体现出详细的规划设计中的眼光。设计需要综合考虑整个湿地处理链，同时要考虑将 5 号湿地作为湿地系统的蓄水池。现浇的高质量混凝土墙形成公园半围合蓄水通道。澳派通过引领潮流的简洁生动的设计，让公园得到功能的提升。

The concrete arc sets both the infrastructural and organisational logic of the park and provides an informal seat and edge to the wetland. Fluorescent lights (which are triggered by a light sensor which responds to the environmental conditions) are housed in the wall creating a safe and usable space whilst also being an attractor at night.

湿地旁的弧形混凝土坐墙既成为公园基础设施和组织形式的排列逻辑规则，又为湿地提供了坐处。在混凝土坐墙内存设有光线感应的荧光灯，保证夜晚公园的照明与安全也吸引了人们的眼光。

Design excellence and functional quality:

设计优点和功能质量：

Wetland 5 has been designed to be unashamedly of its times as a contemporary park design. Its materials used, including concrete and timber are robust and appropriate for its purpose.

5 号湿地的设计非常简约现代当之无愧领先时代。其选用的材料，如混凝土和木材等，都非常适合设计目标。

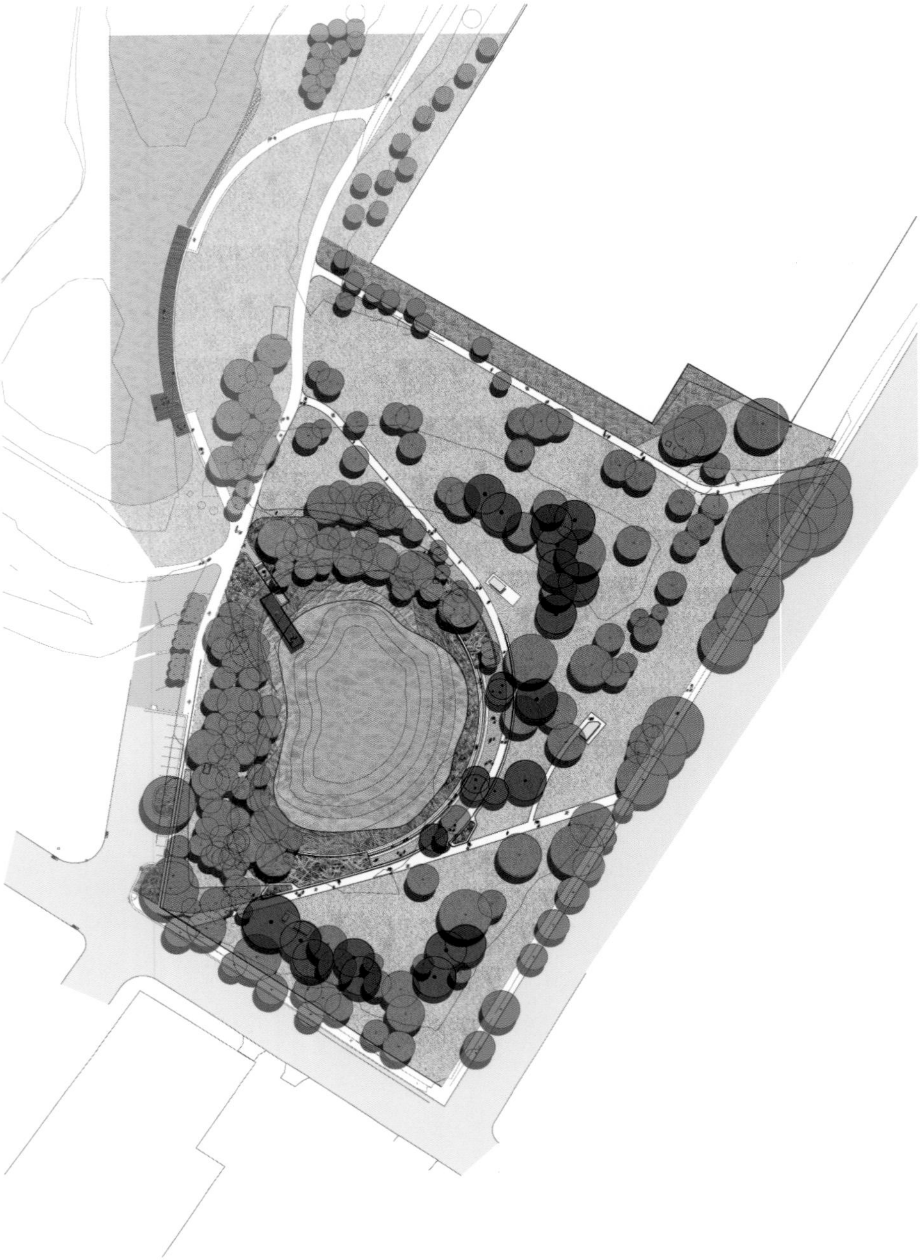

The design does not rely on traditional wetland motives and materials to reveal its function such as the often seen sandstone boulders and decorative artworks. The bold off-form concrete arc, which defines the extent of the wetland, relates to the greater industrial context of the St Peters area and the "industrial sublime" of Sydney Park.

设计并不依赖于传统的人工湿地样式和材料来显示其功能，如传统湿地常见的大石块和装饰艺术品。设计采用大手笔的设计方法，采用一道弧形的混凝土坐墙，确定了湿地的范围来反映圣彼得斯大工业区和悉尼公园的工业风格。

The contrast of the concrete planters and wall with the fine-textured wetland plants and fauna that they attract creates a rich play. The simple gesture of the arc is the key to the design and has been maintained through the detailed master plan to construction. The arc is a clear diagram which terminates the watercourse through Sydney Park and provides bench seating along its entire length.

弧形混凝土墙的简洁与直线的景观与公园的自然景观形成鲜明的对比，给人们一种强烈的视觉冲击感。简洁的弧形是设计的关键，这个设计保证从图纸到施工得到了贯彻实施。弧形混凝土墙对于湿地水体的流通起到一个导向作用，而且整个坐墙都可以供人们休息。

As this was an insertion and a upgrade to an existing park, there was a range of constraints which were sensitively dealt with. These included:
preserving existing trees;
maintaining opportunities for existing park users by maintaining and upgrading desire lines;
resolving the hydraulic engineering issues of the wetland as well as the obvious expectation to greatly improve on the existing degraded open space.

由于湿地设计是在现有公园基础上建造的，存在许多限制条件，需要谨慎处理。这些措施包括：
保留现有乔木；
保留公园的功能并提升公园的功能；
解决湿地的水利工程问题，改善现有的公共空间环境。

The upgrade to Wetland 5 has provided a significantly improved wetland and park experience as well as improving safety and surveillance opportunities for passive recreation.

5 号湿地的改造不仅提升了整个湿地系统的功能，而且还大大提高了公园静态活动空间的安全性，便于公园的管理。

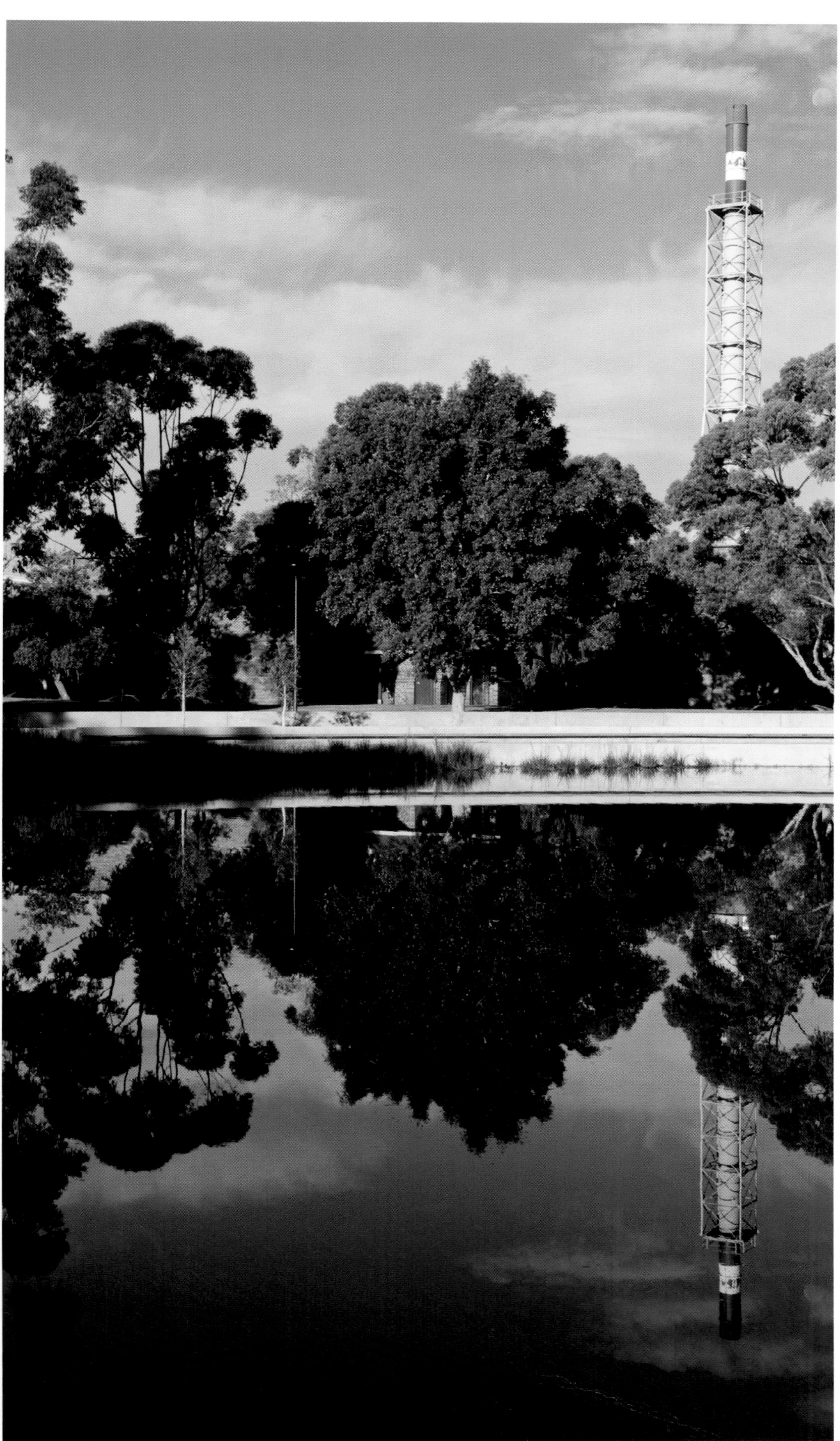

Environmental responsibility and sustainability: A wetland project by its nature alone does not simply comply with the criteria for excellence in Environmental design, though it is a good start. Wetland 5 was a wetland prior to this upgrade albeit degraded and loosing water through breaks in the substructure.
保障生态可持续发展，对自然环境负责：湿地设计并不只是追求生态的可持续发展，但是设计的开端考虑生态的功能是很好的。5 号湿地在改造前就是一个湿地，但是由于结构上存在问题造成湿地退化和水的流失。

Wetland 5 is the key to the entire wetland system as it is in the lowest part of the park. The water is gathered here and pumped back through the system. The wetlands have made significant new growth in an extremely challenging growing medium (a thin crust of soil over a rubbish tip) and as a consequence have created new eco systems which attract biodiversity (birds and other fauna) to live in this semi industrial park.

5 号湿地是整个湿地生态系统的关键，因为它位于公园地势的最低点。所有湿地处理后的水体都聚集在 5 号湿地，然后通过水泵把水体抽到地势最高点，再进行新的水体循环。湿地使新的生命能够在极具挑战性的环境中（垃圾上方的薄土层）成长，创造新的生态系统，吸引多样性生物（禽鸟和其他动物）到此栖息。

Wetland 5 is a contemporary example of the "landscape cyborg". It is a contemporary urban park underlain with significant environmental design—much of it is hidden to the eye. Wetland 5 is a relevant example of the fusion of environmental objectives and high-quality design that creates memorable and enduring parkland.

5 号湿地的景观是一个现代简约设计的成功范例。湿地不仅是一个现代城市公园，而且还考虑了大量的环境设计——尽管人们可能看不到这些环境的设计。5 号湿地融合了环境设计与精致的景观设计，打造一个令人难忘和长久的公园空间。

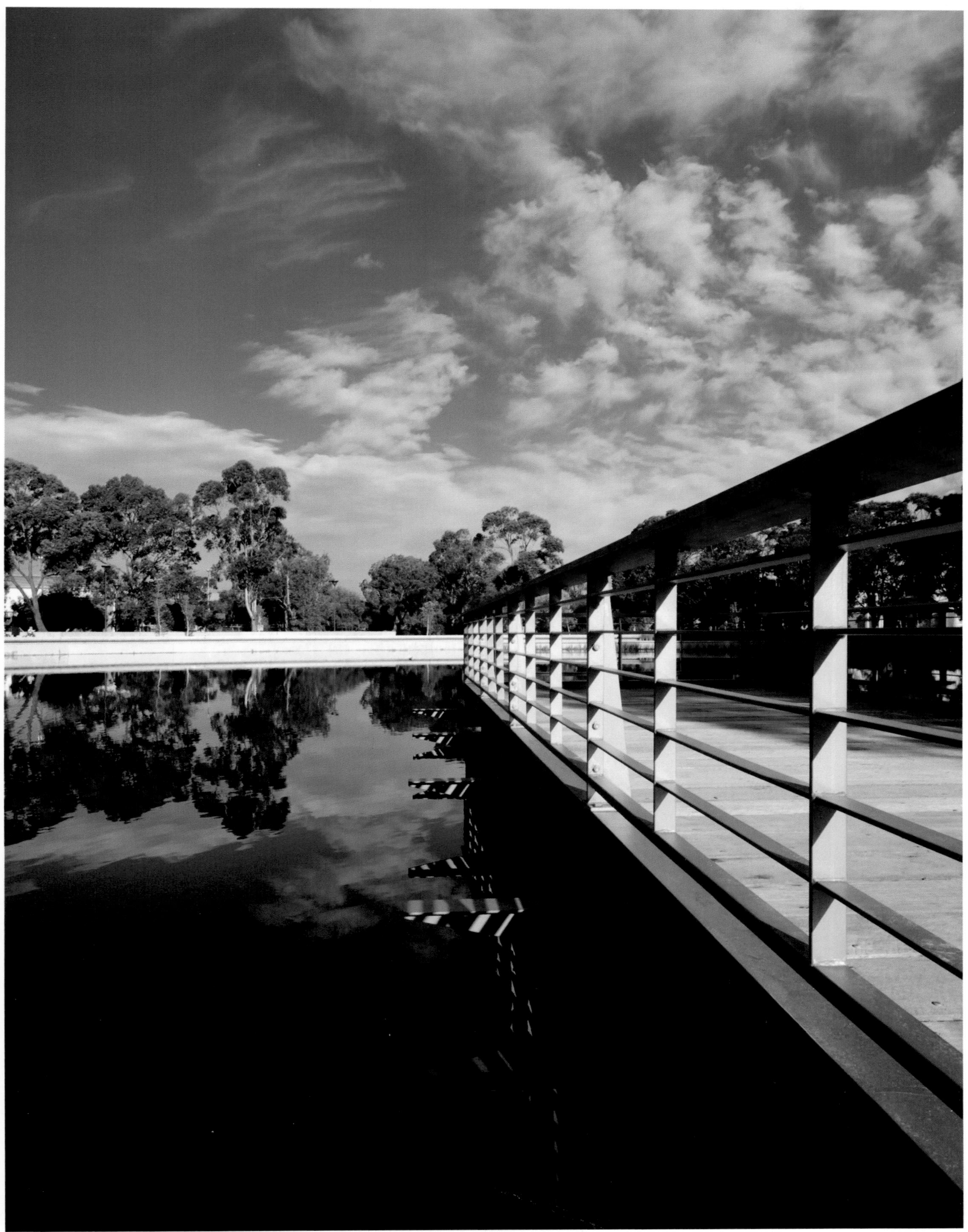

Bondi to Bronte Coast Walk Extension Design Statement

Bondi 至 Bronte 滨海走廊延伸段景观设计

LOCATION：Sydney，Austrilia
项目地点：澳大利亚 悉尼

AREA：200,000 m²
面积：200 000 平方米

PHOTOGRAPHER：Florian Groehn
摄影师：Florian Groehn

DESIGN COMPANY：ASPECT Studios
设计公司：澳派（澳大利亚）景观规划设计工作室

Bondi to Bronte Coast Walk Extension Design Statement

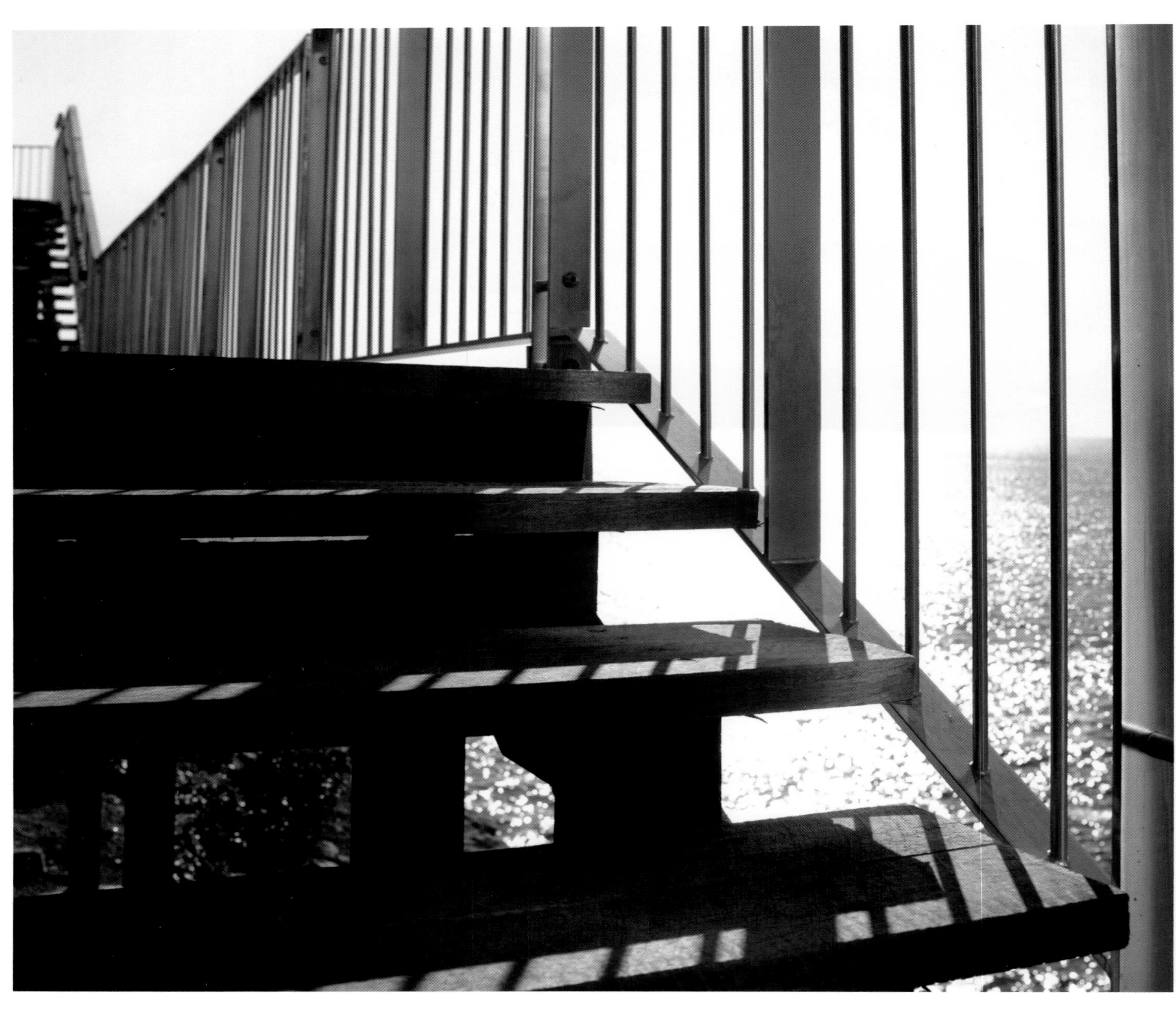

Sydney's sublime headlands, sandstone outcrops and stunning regional views are expressed along this environmentally sensitive elevated boardwalk high above Sydney' s Eastern cliff tops.

站在澳大利亚悉尼东边悬崖顶环境敏感的大道上，悉尼壮丽的海岬景观，砂岩矿脉和美轮美奂的风景尽收眼底。

This 515 m long walkway is part of the nationally significant 9 km coastal walk from South head to Maroubra. The project resolves complex geotechnical, structural and heritage conditions and retains the significant cliff-top heath community on the hanging swamps along the exposed sandstone platforms. As a direct result, the materials of the boardwalk change from timber to a gridded fiberglass, allowing light and water to go through remnant vegetation.

澳派的设计在保留周边的文化遗产的同时，为悬崖峭壁处的观景创造了独特的体验。515 米长的滨海漫步道是全澳大利亚闻名的 9 000 米滨海景观漫步大道的一部分。项目解决了复杂的地质工程和结构问题，同时保留了裸露砂岩平台上的湿地分布的壮观的崖顶社区。大道的材质也由木平台变成玻璃纤维土工格栅，保证走廊下方的植物能够容易地吸收阳光和雨水，从而保护了植被。

Five Lookout points with bespoke furniture create opportunities to pause, rest and enjoy the spectacular views along the sandstone coast.

滨海走廊设有五处观景台，设有定制的座椅小品，充分考虑到游客的驻足休憩，享受砂岩海岸的独特美景。

The walk is strategic about the balance of "risk" to "experience" and as such the balustrades are limited to areas which most require safety.

项目使用了简单的设计语言，整个滨海走廊设在悬崖上方，在保证安全的前提下让行人充分享受一路的风光。

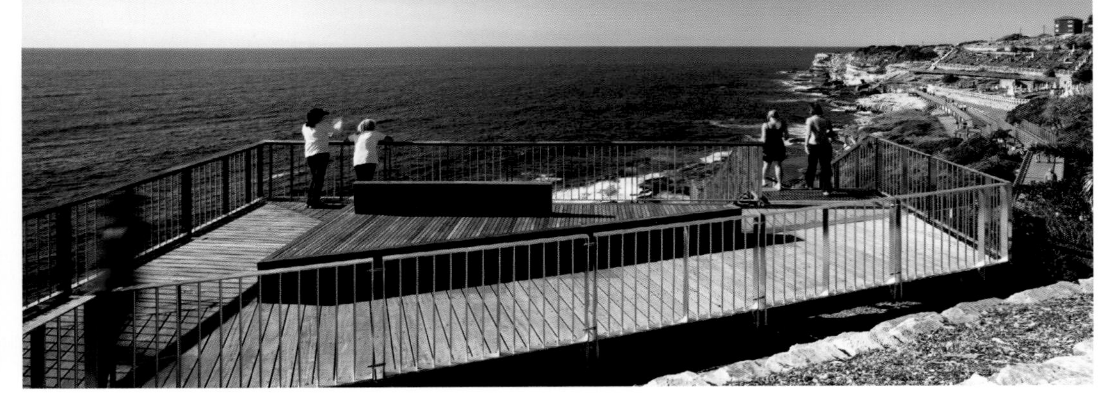

The project has a clear design language, using simple materials. It shifts and slides along the movement path to reveal the story of the cliff–top landscape—including hanging swamps, exposed rock outcrops and rich habitat ecology. The walkway cranks and fractures at the lookout points in direct response to the crystalline geology of the site and angular structure of Hawkesbury sandstone. The lightness and calibration of the structure which moves from ground to an elevation of 7 m in height reinforce this experience of being on the edge.

整个滨海走廊采取清晰统一的设计语言，选择简洁耐用的材料。行走的路线蜿蜒曲折，逐步向人们展示出悬崖顶的景观——沼泽地、岩石矿脉和丰富的生态景观。观景台的设置与 L 形或折线形的步道相结合，步道从地面上升 7 米，其轻便和标准进一步增强了在悬崖边上观景的效果与体验。

The site planning was rigorous and determined by ecological (avoiding remnant vegetation), geotechnical (rock outcrops, unstable fill slopes, dyke intrusions) and public safety (avoiding potential for cemetery wall collapse).

严格的设计规划中仔细考虑了生态走廊（尽量不去干扰现有的植被）、岩土（矿脉、不稳定的山坡、堤防）和公共安全（避免墙体倒塌）。

All materials selected are robust, long–lasting and are selected to minimize ongoing maintenance by the Council. A combination of materials was used for the decking to achieve aesthetic, ecological and durability goals. Fiberglass mesh was used over areas of important remnant vegetation to allow some light penetration, rain permeability and for its construction ease and durability. It was decided that due to the extreme coastal exposure, fiberglass would be the only material to withstand corrosion and meet the design quality and aesthetic objectives. The stainless steel balustrades were electro–polished to reduce "tea staining" corrosion.

在材质的选择上，选用了坚固、持久性强的材料，以减少后期的维护成本。观景平台选用多种材料以期达到美观、环保和持久的效果。在多处的残存植被处，使用玻璃纤维格栅，保证滨海走廊下方的植物可以自然吸收阳光与雨水，同时也降低了施工的难度，保证使用的持久性。鉴于海岸边的自然气候环境特殊，玻璃纤维是最佳的材料，能经受腐蚀，能达到质量要求并具有美观效果。不锈钢的扶手经过电抛光，杜绝油斑腐蚀。

Frankston Foreshore Precinct

Frankston 滨海区

LOCATION：Melbourne，Australia
项目地点：澳大利亚 墨尔本

AREA：300,000 m²
面积：300 000 平方米

PARTY A：Frankston Municipal Council
甲方：Frankston 市议会

TYPE：Urban Public Space Design/Ecological Design
项目类型：城市公共空间设计／生态设计

SCOPE：Master Planning，Conceptual Planning，Deepening Design and Constructional Drawing Design
设计范围：总体规划、概念规划、深化设计及施工图设计

COMPLETION DATE：2006
完成时间：2006 年

Frankston Foreshore Precinct

DESIGN COMPANY：ASPECT Studios
设计公司：澳派（澳大利亚）景观规划设计工作室

The Frankston Foreshore Precinct is a large civic space and car park ,It connects the commercial precinct with the natural assets of the foreshore. It provides a recreation focus for locals and visitors to Frankston.

Frankston 滨海区是一个大型的市民活动空间和停车场，连接商业区与自然滨海区，是一个集中了当地居民和游客的娱乐中心。

Most coastal developments segregate activity in hierarchy of use, parallel to the beach. The key design intent of the precinct is to break this hierarchy, realigning the space perpendicular to the beach. This is achieved through a robust underlying structure, occupied by various activities, surface areas, art works, the Landmark Bridge over Kananook Creek and vegetation types.

设计打破了通常滨海景观与海滩平行按使用等级分隔活动的模式，通过各种不同的活动的设置，铺地形式，艺术品，跨越 Kananook 河的地标性的景观桥以及植物配置，打破这种等级，重新调整与海滩垂直的景观带，将人们引导到滨海区域。

The precinct is significant as it uses public open space as the main catalyst to align the city to the Bay. This was not its original brief but through the master planning and design process ASPECT sought to grasp this opportunity. Now the precinct is a public space commanded by the bay and car parking is removed as the main public gesture. Also the bridge has become a landmark of Frankston and is a symbol for its revitalization.

本项目的意义重大，主要通过打造滨海公共空间，将城市引向海边。这本来不是最初设计任务书的要求，然后，在总体规划设计的过程中，澳派抓住了这个机会，将项目现场设计为滨海公共空间，并将停车场移到外围区域。此外，景观桥也成为 Frankston 的城市地标，是城市重新繁荣的象征。

It marks the beginning of a series of projects that will ensure a high-quality public foreshore domain for Frankston.

本项目标志着将 Frankston 城市打造为高品质的公共滨海区的一系列项目的开端。

RAINER SCHMIDT LANDSCAPE ARCHITECTS

OFFICE PROFILE

With over 20 years of professional work experience as a landscape designers Rainer Schmidt Landscape Architects is one of the largest and successful offices in Germany.

The office has roughly 30 multi-national staff members including senior landscape architects, landscape architects, graphic designers and administrators. The company has three locations in Germany: The main office is in Munich, the others in Berlin and in Bernburg. An office in Beijing manages some of the Asian projects. An office in Turkey and Dubai will follow in the near future due to the increase of projects in these regions.

Prof. Schmidt is a professor at the University of Applied Sciences in Berlin and adjunct professor at Peking University.
The scope of projects and working fields of the office are the planning and execution of large projects in the fields of landscape architecture, environmental planning, urban design and supervision on both national and international level.

The company´s aim is to find ways of dealing with problems of our time. The language of landscape architecture in the 21st century must offer a realistic reflection of the way people interact with each other and with nature. The office is striving to achieve a balance between design, function, emotion and conservation.

Villa H. St. Gilgen

H 别墅，圣基尔根

LOCATION：St．Gilgen，Austria
项目地点：奥地利，圣基尔根

AREA：2,900 m^2
面积：2 900 平方米

DESIGN COMPANY：Rainer Schmidt Landscape Architects
设计公司：赖纳·施密特景观建筑公司

Villa H. St. Gilgen

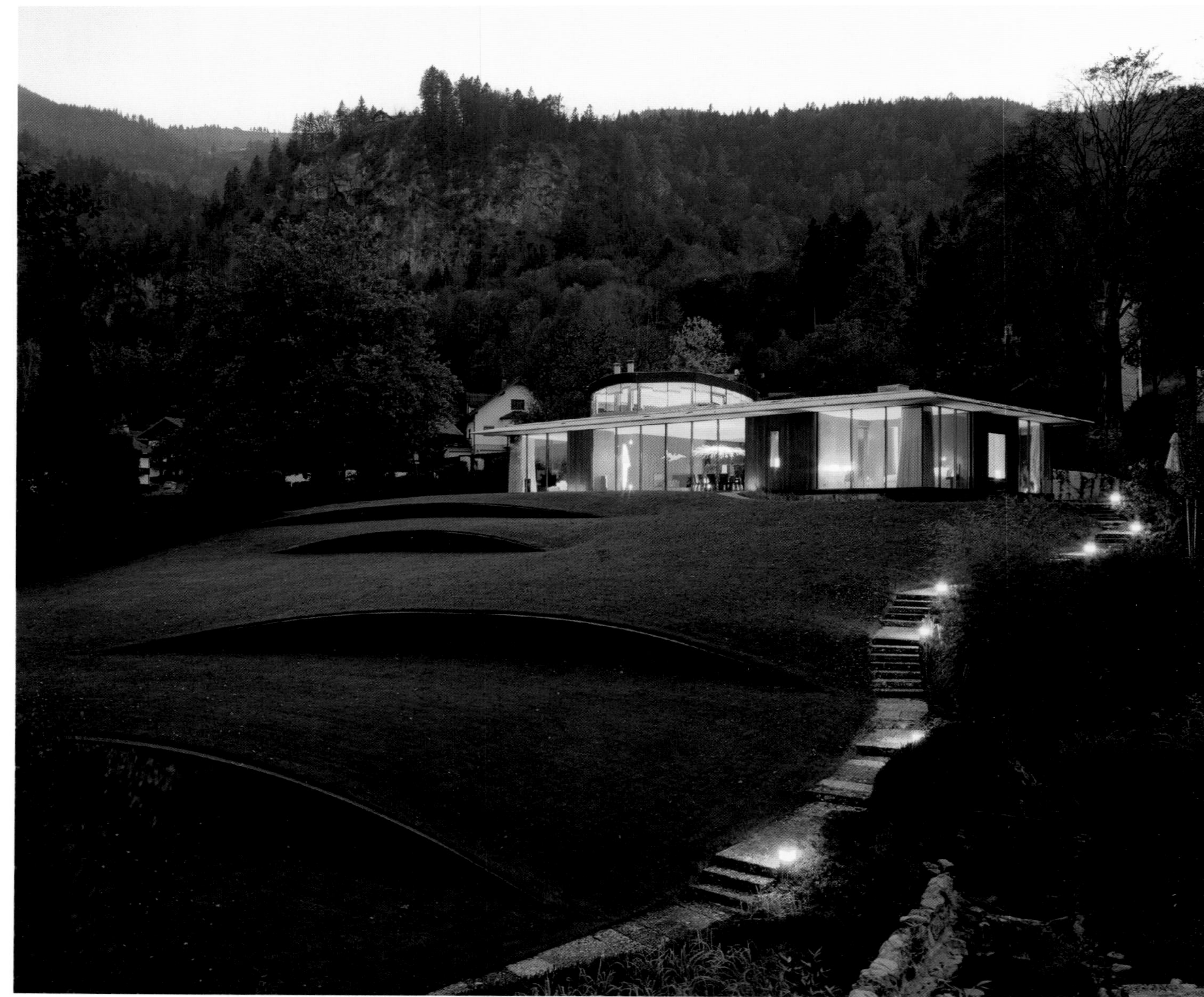

An undulating landscape is the new connection between the new and old living quarters of a villa in St. Gilgen near Wolfgang Lake. The calm lawned area accentuates the huge Alps and lake view the panorama of.
The modeled grass area is bordered by a small stream as well as roses and perennials. Corten steel arches like sickles cut through the green lawn.
The "sickles" change their colour depending on the weather. The more humid weather makes them look almost black whilst in the sun they have a bright orange red glow. In winter they have white frost.

延绵起伏的景观将位于沃尔夫冈湖附近圣基尔根市的一栋别墅的新旧生活区连接在一起。静谧的草坪区突出了高耸的阿尔卑斯山和湖面景色构成的全景图画。
这片示范性草坪区以一条小溪为边界，点缀着蔷薇花和多年生宿根花卉。柯尔顿钢造的镰刀形拱墙横穿绿色草坪。
"镰刀"的颜色会根据天气的不同而改变。天气越是潮湿，其颜色越暗；在阳光下，它会呈现绚丽夺目的橙红色。冬季则出现白霜色。

The entrance area flowers profusely with Dahlien and Hemerocallis. The pond is planted with blue and red violet hues with Hemerocallis and Iris. Different grasses add to the flower arrangement.
A light-colored grass band grows between the high yew hedge along the road and a low bux hedge.
All roses growing in shade and semi-shade aspects are historical species that have an intense scent.

入口区的花卉以大丽花和大花萱草为主。池塘边栽种蓝色和红紫色调的萱草花和鸢尾花。花间点缀各种草类植物。
沿街道高耸的紫杉树篱与低矮的黄杨树篱之间种植浅色草。
种植在阴凉和半阴凉处的蔷薇花都是浓香的历史物种。

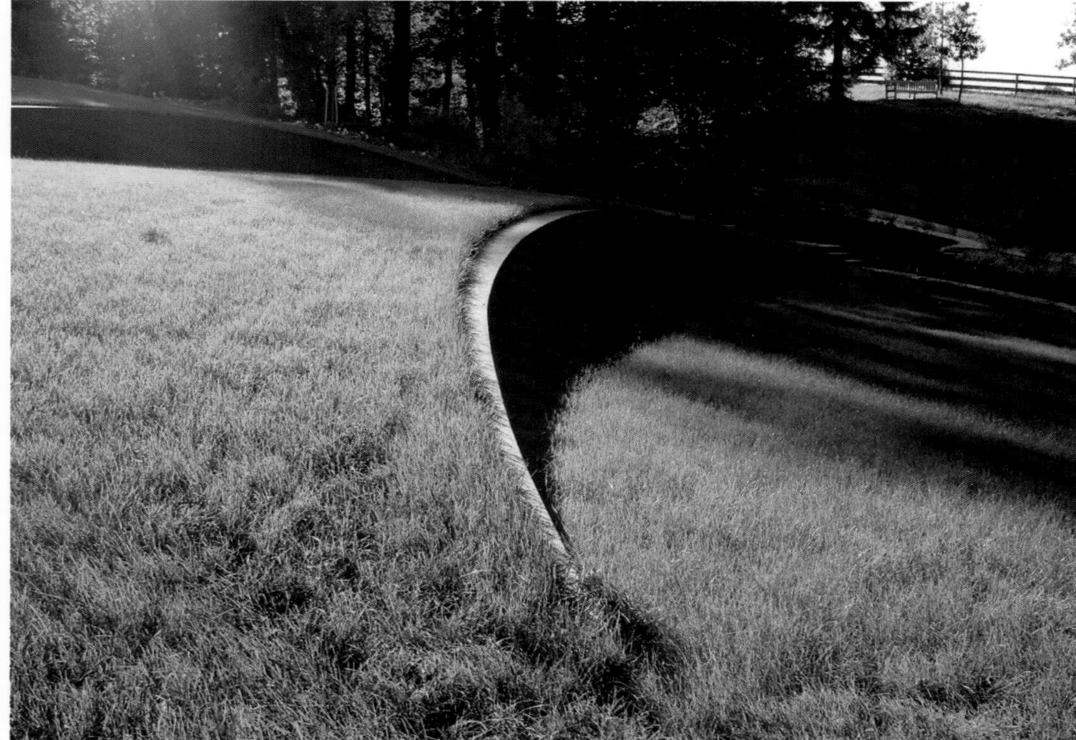

The garden thrives on the strong contrasts such as the calm lawn and the wild and colorful perennial planting. Single-seating areas are positioned with different characters depending on the daily preferences: Sunny overlooking the lawn, shaded under the trees or amongst the sweet-smelling roses.

在静谧草坪和野生植物及鲜艳的多年生植物强烈对比之下，花园显得繁花茂盛。单座椅区根据日常选择被赋予了不同的特点：瞭望草坪处阳光灿烂，树荫下或甜蜜花间则荫蔽幽暗。

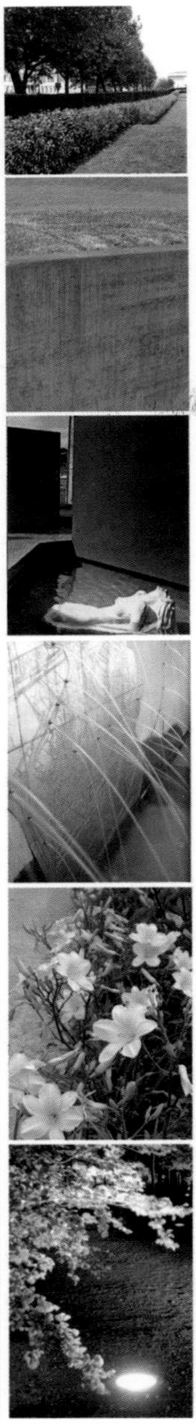

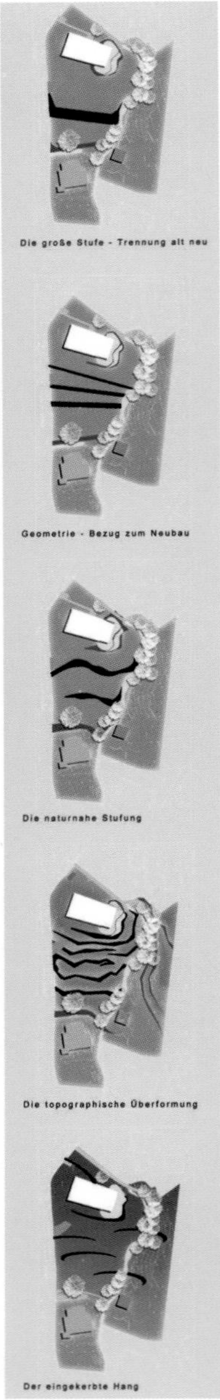

Die große Stufe - Trennung alt neu

Geometrie - Bezug zum Neubau

Die naturnahe Stufung

Die topographische Überformung

Der eingekerbte Hang

Villa Garden Bogenhausen

克兰茨别墅

LOCATION：Munich，Germany
项目地点：德国 慕尼黑

AREA：3,000 m²
面积：3 000 平方米

COMPLETION DATE：2006
完成时间：2006 年

DESIGN COMPANY：Rainer Schmidt Landscape Architects
设计公司：赖纳·施密特景观建筑公司

Villa Garden Bogenhausen

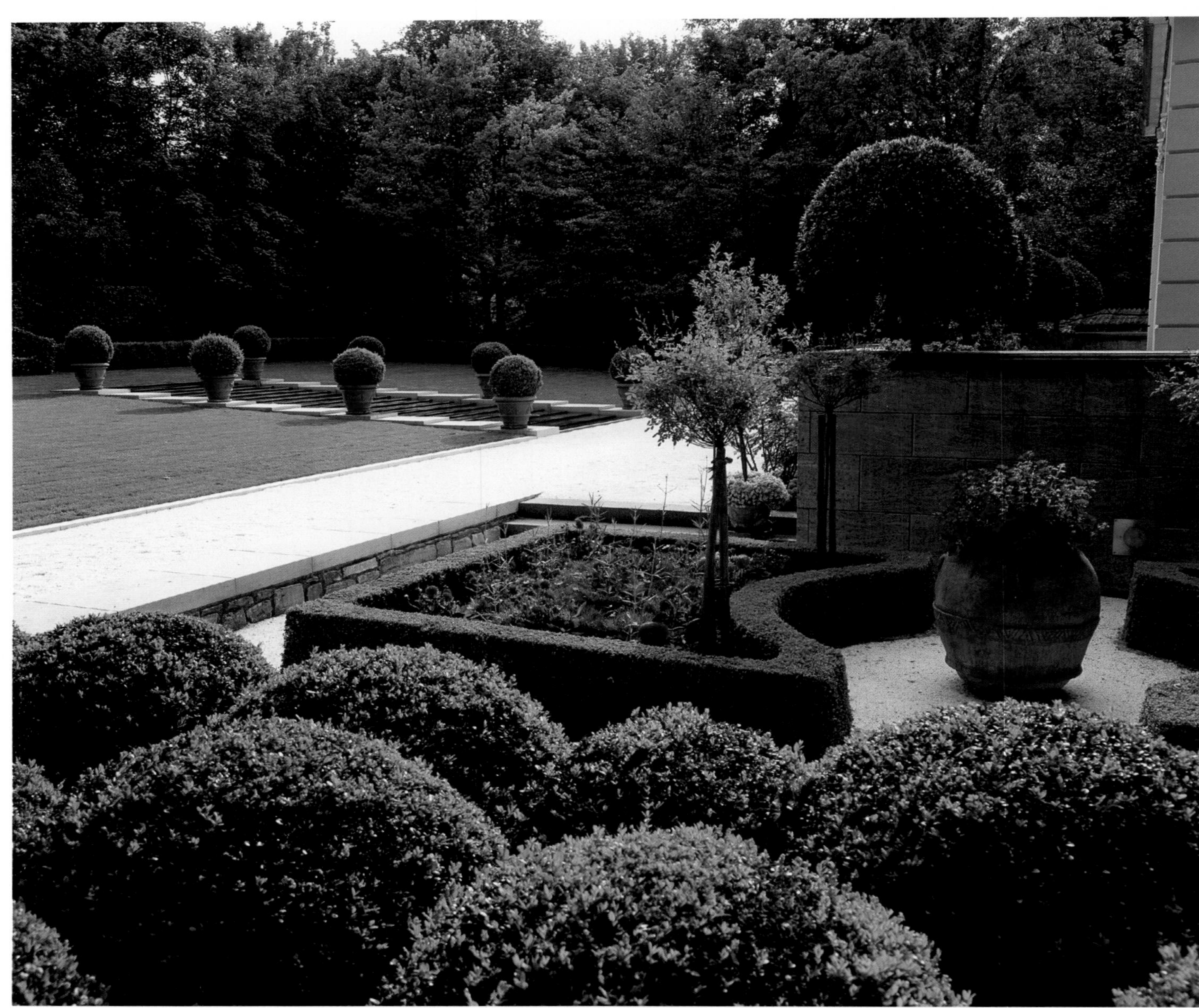

The original villa was built with a representative Neo–Classical style in 1923 and refurbished in 1980. Clear lines and a reduced architectural language dominate the garden character that is framed with a wild "nature". Tasteful and high–quality materials such as natural stone and bronze strengthen the clear concept.

别墅始建于1923年，具有典型的新古典主义风格，并于1980年进行了修缮改建。整个别墅花园的外部充满野趣的“自然”气息，内部则通过简洁明快的线条和简练的建筑语言来表达。各种高雅、高档材质的使用更加强化了这一设计理念，如自然石材以及青铜雕塑的使用等。

One of the three areas is the main entrance east of the villa. The entrance is bordered by over–sized bux hedge and rhododendron. The second area is a bux parterre right and left of the terrace planted with a summer theme. The third and largest area is a field of rising grass dissected with cascading water steps that are framed with cut bux hedges in terracotta pots. An ivy settee, with various trimmed hedges of different greens and textures as a background, forms the focal point at the end of the cascade. This lawn area is raised and supported by stone–clad retaining walls.

花园主要分成三个部分，其中东面是别墅的主入口，入口处种植着大片的黄杨和杜鹃花丛。第二个部分是平台左右两侧以夏季为主题栽种的黄杨花坛。第三部分最大，是一片高出地面的草地，草地上有流水台阶，与赤褐色花盆中的黄杨绿篱相互映衬。一条常春藤靠背长椅，衬以修剪整齐的各类绿色植物组成树篱和肌理，形成了瀑布尽头的焦点。这片草坪区是突起的，周围用石质挡土墙支护。

An unusual perspective is created by the graded grass, different hedges and the water–feature view. Together the elements rise and form a symphony of fresh green, summer flair and cooling waters.

斜坡草坪、形态各异的树篱和水景创建了不同寻常的景象。这些元素共同构成了由清新绿地、夏季情调和凉爽水域谱写的交响乐。

Weser Quartier Bremen

不来梅 威悉小区

LOCATION：Bremen，Germany
项目地点：德国 不来梅

AREA：9,800 m²
面积：9 800 平方米

COOPERATION PARTNER：Murphy Jahn Architects，Berlin，Chigago
合作伙伴：芝加哥墨菲·扬建筑事务所柏林分所

PLANNING：2008–2010
规划：2008–2010 年

DESIGN COMPANY：Rainer Schmidt Landscape Architects
设计公司：赖纳·施密特景观建筑公司

Weser Quartier Bremen

The Weser Quarter is being built at the entrance to the "berseestadt" of Bremen, directly on the Weser River promenade within walking distance of the city center. The grounds add to the coherence of the different building typologies and uses with their unified structure. On the side toward the water a terraced landscape with steps invites the visitor to take a sunbath. The terraces can be used as lounges by the neighbouring restaurants and cafés or as a grandstand for events.

威悉小区位于不来梅港口的入口处，直接建在威悉河散步区之上，步行即可到达市中心。小区协调了不同种类的建筑，采用了统一的建筑结构。在朝向港口的一边有一个台地景观，游客可以在这里享受日光浴。台地还可以用作周围餐厅和咖啡馆的休息区，或在举行大型活动时用做大看台。

ANDREA COCHRAN LANDSCAPE ARCHITECTURE

OFFICE PROFILE

With over 20 years of professional work experience as a landscape designers Rainer Schmidt Landscape Architects is one of the largest and successful offices in Germany.

The office has roughly 30 multi-national staff members including senior landscape architects, landscape architects, graphic designers and administrators. The company has three locations in Germany: The main office is in Munich, the others in Berlin and in Bernburg. An office in Beijing manages some of the Asian projects. An office in Turkey and Dubai will follow in the near future due to the increase of projects in these regions.

Prof. Schmidt is a professor at the University of Applied Sciences in Berlin and adjunct professor at Peking University.
The scope of projects and working fields of the office are the planning and execution of large projects in the fields of landscape architecture, environmental planning, urban design and supervision on both national and international level.

The company´s aim is to find ways of dealing with problems of our time. The language of landscape architecture in the 21st century must offer a realistic reflection of the way people interact with each other and with nature. The office is striving to achieve a balance between design, function, emotion and conservation.

Peninsula Residence

半岛公馆

LOCATION: California, USA
项目地点：美国 加利福尼亚

LEAD DESIGNER: Andrea Cochran, FASLA
首席设计师：Andrea Cochran, FASLA

DESIGN COMPANY: Andrea Cochran Landscape Architecture
设计公司：Andrea Cochran Landscape Architecture

Peninsula Residence

Situated on a suburban street, the Peninsula Residence aims at creating an alternative model for the standard front yard that is appropriate to neighborhood and the Northern California climate. The landscape design re-imagines a previously neglected slope to implement sustainable materials and an interconnected spatial relationship between public and private zones. This project exemplifies how the innovative reconfiguration of an entry can create a place that engages a home, a street, and the senses. In the surrounding neighborhood, two primary approaches are taken to address the yard. One creates an open lawn area, offering no mediation between the house and the street and requiring significant water and resource investment. The other approach prioritizes privacy of the domestic realm, creating a tall opaque barrier between the house and the street either with thick hedging or a privacy fence. For this 1/3 acre renovation, the design intent was to create a landscape of private refuge while allowing the front of the house to remain connected to the neighborhood. It carves out a series of intimate, interconnected rooms from a previously unused band of land in front of the house. The landscape architects collaborated closely with the architecture team to reshape space, redirect views and develop a protracted threshold sequence. Prior to the redesign, the entry consisted of a direct, intimidating stair, leading up a steep slope, straight to the front door. The slope offered no usable space for the residents in the front of their house. Rather, this space was planted with an assortment of ornamental plants with high water use and no overall coherence. The clients, who had lived in the house for two decades, were taking the opportunity to renovate the house and garden to suit their changing needs. One powerful aspect of the existing landscape, a grove of native oak trees, planted by the current owners when they first moved to the property, was recognized by both landscape architects and clients as the most significant expression of the place's identity.

位于郊区临街的半岛公馆，旨在创建一种可替代常规前院的模式——适于附近地区及北加利福尼亚温暖晴朗气候的住宅区。景观设计采用可持续性材料，使室内和外界形成一种空间互联的关系，重新定位了一个以前被忽略的边坡。玄关的革新重构，不仅愉悦感官，而且使居所和街区融洽衔接，半岛公馆正是例证了这一点。有两个途径可以从周边社区通向院子。一是只通过一块开放的草坪，在房屋和街道之间不做任何修整，需要投入大量的水和资源。第二种方式则优先考虑居室的私密性，用树篱或栅栏创造一个高且不透明的的屏障将其与街道界开。对这 1/3 英亩的设计，不仅仅是要建立一个私人景观居所，同时还要使其看起来与周围融洽。将居所前那块原来空闲的地方，营造成为一个温馨的私人空间。景观设计师通过和建筑设计团队的紧密合作，重塑空间，改变景观，修建了这个繁复的玄关。重新设计前，入口包括一个径直的、岌岌可危的楼梯，坡度很陡，一直延伸到前门儿。这陡坡使居住者无法利用房前的空间，然而，这儿却种植着各式各样的植物，不仅消耗大量水源，而且与空间整体也并不那么协调。在这居住了二十年的业主，也利用这个机会来改建他们的房屋和花园，以迎合他们不断改变的需求。原有景观非常重要的一方面是一片橡树林，是业主刚入迁时所种，景观设计师和业主都认为他们是这块区域最好的身份象征。

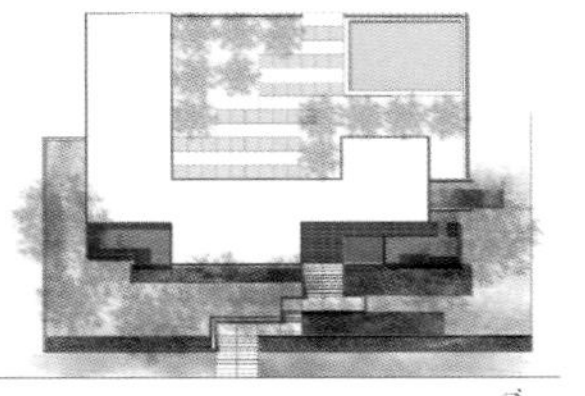

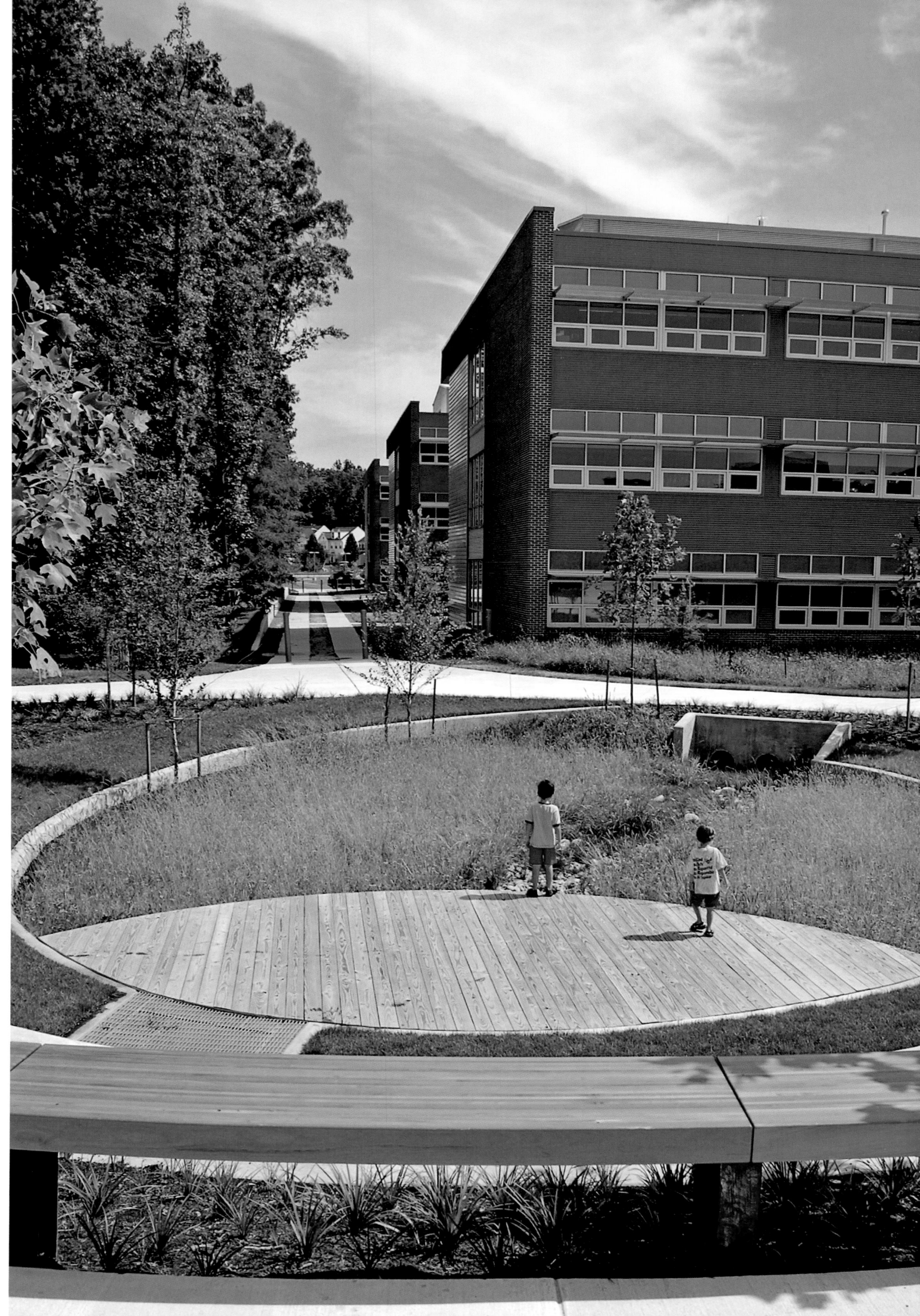

Iteworks Studio

OFFICE PROFILE

Siteworks is an award-winning landscape architecture studio based in Charlottesville, Virginia. We are a small, dynamic studio dedicated to making meaningful human places that intelligently engage local ecologies, are culturally relevant and that operate simultaneously as habitat and as living sculpture. We are involved in projects ranging from gallery installations, site sculpture and private gardens, to corporate and institutional landscapes, post-industrial sites, urban design and public parks. We are expert collaborators and are sought after by architects and engineers to add value and creativity to integrated design teams on projects throughout the U.S. and in Central America, Canada and Europe.

In our practice we believe that all design must adhere to principles that protect and enhance watersheds, improve the ecology of sites, increase biodiversity and carefully manage the impacts that human processes have on a larger context. Through the development of high performance projects we aim to go beyond sustainability and towards regenerative and restorative solutions. We have proven ability in the artful integration of human systems and native ecologies.

We believe that it is vital for the design of human places to be beautiful, functionally effective and intellectually stimulating. In all of our projects we work toward a fusion of ecology and culture that recognizes the role of human society in a larger environmental context. We do not operate from a formal ideology, but rather from one that is based in a careful analysis of the sites and people with whom we work. Our process employs dialogue and creative expression to fuse the needs and desires of our clients with the inherent qualities and conditions of the places they live.

Manassas Park Elementary School Landscape

马纳萨斯公园小学

LOCATION：Virginia，USA
项目地点：美国 弗吉尼亚州

LANDSCAPE ARCHITECT：Siteworks
景观设计师：Siteworks

TEAM：Nelson Byrd Woltz Landscape
团队：Nelson Byrd Woltz Landscape

DESIGN COMPANY：Iteworks Studio.
设计公司：Iteworks Studio.

Manassas Park Elementary School Landscape

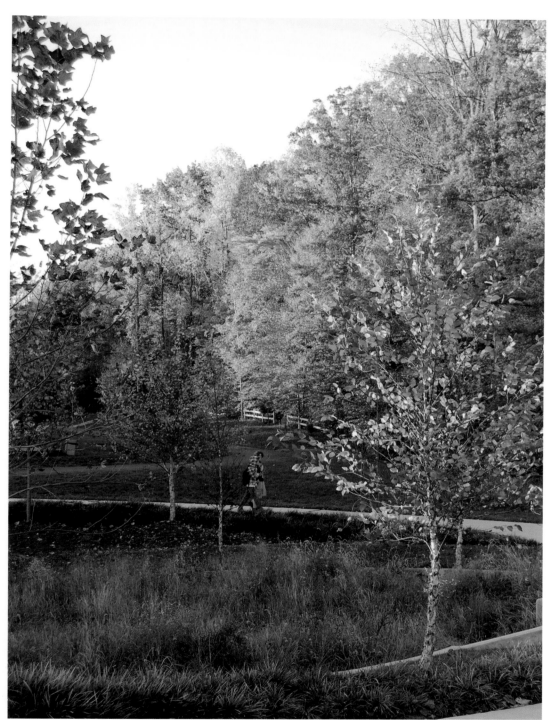

Studies have shown for years the benefits to students' performance and health when schools are designed with fresh air, natural daylight and connections to the outdoors. MPES achieves those benefits while assuming the added responsibility of cultivating environmental stewards in their community of teachers, learners and parents.

多年的研究表明，如果学校的空气清新，自然采光，设计多与户外相连接的话，对学生的表现和健康是非常有好处的。而负责给学生及其家长提供这样的社会环境的，是培养他们的教师。

This LEED Gold certified project goes beyond sustainability checklists to create a school that challenges accepted paradigms in teaching and learning by actively involving the entire community in the design and ongoing operations of their school campus. Like Maurice Sendak's Where the Wild Things Are, MPES elicits a sense of imagination, discovery and wonderment through the relationship it creates between a human community and the surrounding woodlands and watersheds within which it lives.

这项 LEED 金奖认证项目超越了那些所谓的可持续性计划，使学校接受那些教学规范的挑战，使整个社会积极参与到校园的设计和日常活动中来。 像是 Maurice Sendak 的《野生物在哪里》，MPES 即通过创造人类社会和周边物种的生存环境的联系，激发想象、发现和惊奇的感受。

Manassas Park, Virginia, is a small, independent city surrounded by the affluent northern Virginia suburbs of Washington D.C. Incorporated in 1975, the city cobbled together a series of pre-manufactured mobile buildings to create its first generation of school facilities from scratch. Ten years ago, the city began rebuilding all of its public schools—an enormous challenge in a city with an extremely low tax base.

马纳萨斯公园在弗吉尼亚州，这是一个独立的小城市，为弗吉尼亚州北部繁荣的华盛顿郊区所环绕，它在 1975 年就为整个州的学校制造了第一代可移动的建筑设施。10 年前，这个城市开始重建所有的公立学校，这对于低税基城市来说，是个巨大的挑战。

The campus sits tightly surrounded by tract housing, private forest, and the historic landmark Camp Carondelet—forested winter quarters of the Confederacy's Louisiana Brigade between the first and second Manassas campaigns.

校园道路两旁是房屋和私有林，还有历史性地标 carondelet 营地——第一次和第二次马纳萨斯运动之间，联邦的路易斯安那旅队冬季森林茂密的驻扎地。

MPES serves a diverse population of students—many from immigrant families. The 2009—2010 enrollment includes sixty-eight percent non-white and twenty-six percent Limited-English-Proficient children. Forty-four percent receive free or reduced-cost lunches. In the context of this rich diversity, the successful transformation of the school culture testifies to the vision and leadership of the Manassas Park City School's administration.

MPES 有着多元化的学生，大多都是来自移民家庭。2009–2010 年度就招收了 68% 的非白人和 26% 的不能熟练说英语的儿童。44% 可享受免费或低价的午餐。在如此多样化的条件下，学校文化的成功转型，体现了马纳萨斯公园市的愿景，也证明了其管理能力。

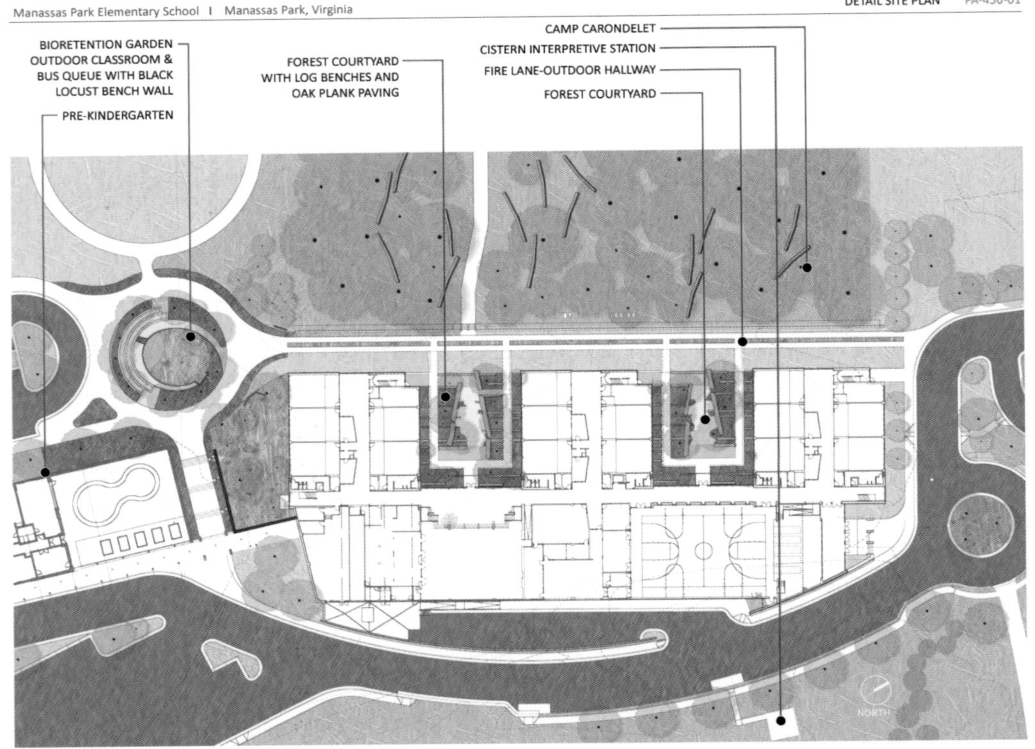

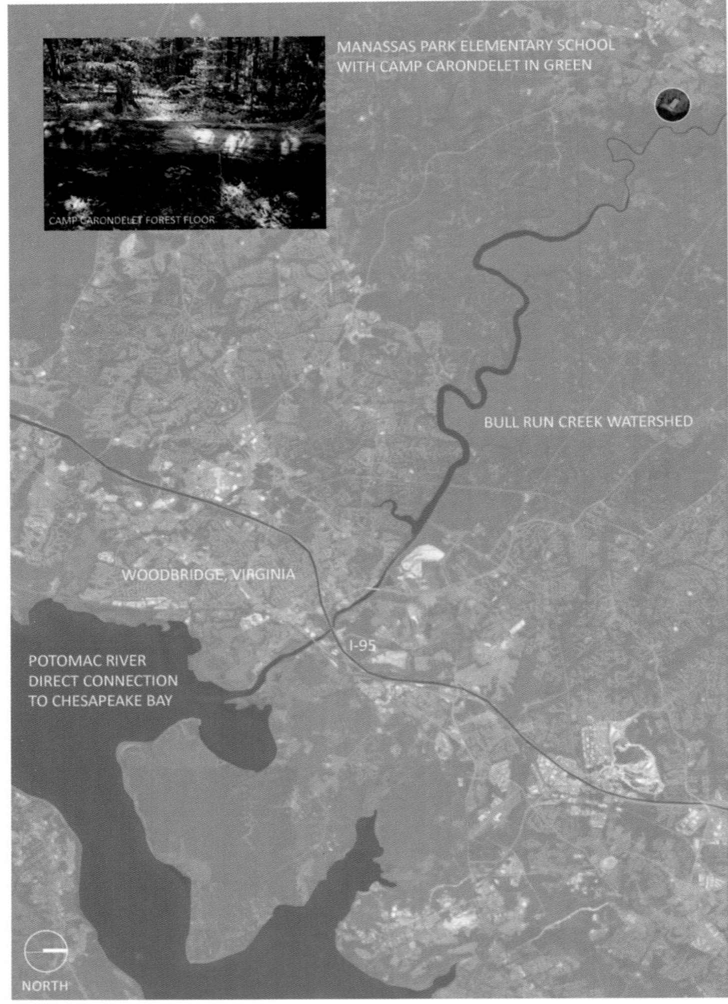

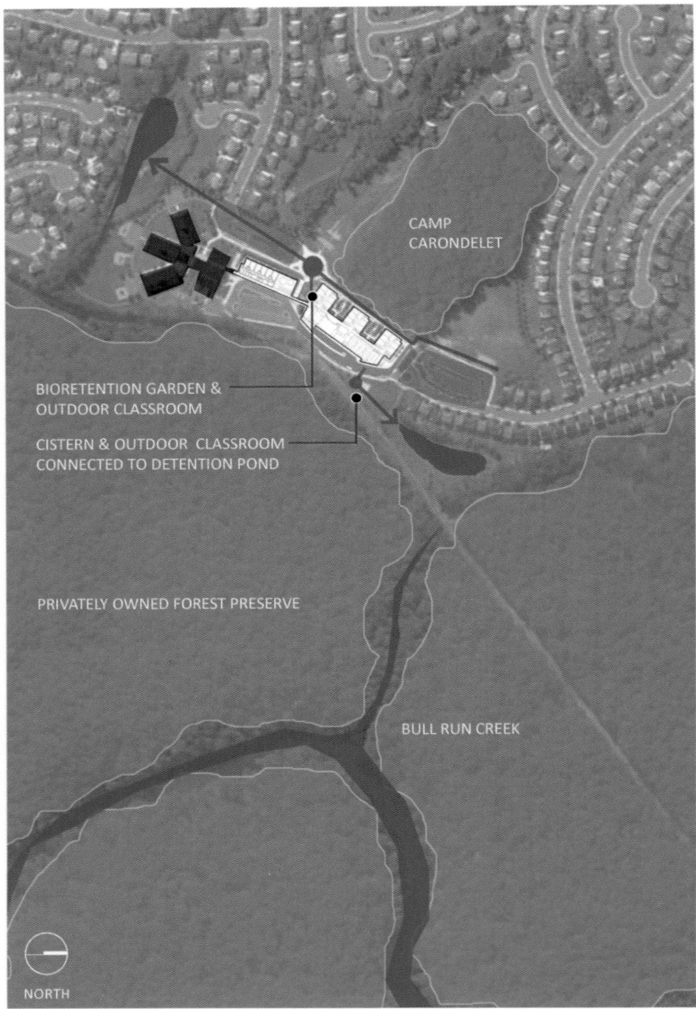

education:

bachelor of landscape architecture &
bachelor of science at Syracuse University

registration:

registered landscape architect
national clarb certification

affiliations:

asla - fellow
aia - honorary member
gsa - national register of peer professionals
nyc parks council
institute of urban design
urban land institute

awards:

asla
bard
tucker
edra/places
institute of urban design
waterfront centre
id magazine
stars of design
aia
nyasla
ny times

THOMAS BALSLEY, Principal Designer

With over 35 years of practice, Thomas Balsley has built a reputation for creating public spaces that enhance and enrich the lives of the individuals and communities who inhabit them. His firm, Thomas Balsley Associates, is an award–winning landscape architecture and urban design firm, whose focus from its inception has been urban parks, plazas and waterfronts.

Mr. Balsley has reshaped urban spaces around the world by designing landscapes that teem with public life and inspire civic pride. His broad and balanced portfolio of award–winning work is respected by both public and private sectors for its design sensitivity and approach to issues of public landscapes. An extraordinary roster of award–winning projects in the public realm attest to his ability to balance the intricacies of the public review process, multiple client interests and challenging budgets without compromise to design excellence.

In the arena of the city, he has faced the dilemmas of craft and art, public service and self–expression that have shaped the profession over the past four decades. In the course of designing scores of urban parks and plazas, he has forged a design philosophy firmly grounded in the dual goals of accommodating the needs of the public while expanding the landscape imagination.

Among his most recent public/private projects are Candlestick Point Park in San Francisco, Curtis Hixon Waterfront Park in Tampa' s downtown, Skyline Park in Denver, West Shore Park in Baltimore and Main Street Garden Park in Downtown Dallas. In New York City alone, Mr. Balsley has completed more than 100 parks and plazas including Chelsea Waterside Park Riverside Park South and Gantry Plaza State Park. In an unprecedented acknowledgement of his contribution to public space in the City of New York, a park on 57th Street was renamed in his honor as Balsley Park.

Mr. Balsley has participated as a panelist and lecturer for the Municipal Art Society, the Institute for Urban Design, the International Downtown Association, the City Parks Alliance and the ASLA. He has also been a guest lecturer and design critic throughout the United States, Canada and the Far East. He was a visiting professor at Harvard' s Graduate School of Design and is an active member of the GSA' s National Register of Peer Professionals.

Every year brings international recognition in the form of awards and citations from professional and civic organizations, including the American Society of Landscape Architects, the Cooper–Hewitt National Design Museum, the American Institute of Architects, Environmental Design Research Association, the Institute for Urban Design and the Waterfront Centre.

Thomas Balsley Associates' work has appeared in national and international publications and media including the recent BBC documentary "Around the World in 80 Gardens," ELA of Korea, the New York Times, the New York Times Magazine, Landscape Architecture Magazine, Sculpture Magazine, Architectural Digest, Abitare, Arredo Urbano, Progressive Architecture, L' Arca, Arredo Urbano, Ville Giardini, Sculpture, International Design, ULI Urban Land, Places, Chinese Architect, and Japan Landscape.

THOMAS BALSLEY ASSOCIATES

Thomas Balsley Associates is an award-winning design firm providing landscape architecture,site planning, and urban design services throughout the United States and abroad.

Over 35years of experience have produced projects of virtually every size and type ranging from master plans to small urban spaces and from feasibility planning studies to completed urban parks,waterfronts, corporate, commercial, institutional, residential, and recreational landscapes.

Thomas Balsley Associates serves a variety of public and private sector clients with a talented staff of professional landscape architects, urban designers and planners who are dedicated to the pursuit of design excellence. The firm's well-earned international reputation is built on a refreshing approach to design and client management in which creativity and innovation are fused with the realities of budget and schedules in a client/designer collaboration.

Gantry Plaza State Park

龙门广场州立公园

LOCATION：New York，USA
项目地点：美国 纽约

AREA：24,281 m²
面积：24 281 平方米

AWARDS：Waterfront Centre，Grand Prize Winner，Tucker Award of Excellence，EDRA/places，Asla Honor Award，Nyasla Honor Award
奖项：滨水中心获得大奖和各种优秀奖，EDRA/ 优秀场所奖，美国景观建筑师协会荣誉奖，美国景观建筑师协会纽约分会荣誉奖

DESIGN COMPANY：Thomas Balsley Associates
设计公司：Thomas Balsley Associates

Gantry Plaza State Park

Once a working waterfront teeming with barges, tugboats, and rail cars, the Hunter's Point shoreline of Queens slowly succumbed to the realities of the post-Industrial Age. As the last rail barge headed into the sunset, this spectacular site was left to deteriorate to a point of community shame. As part of the Queens West Parks Master Plan, Thomas Balsley Associates, together with Weintraub di Domenico, envisioned Gantry Plaza State Park as a place that celebrates its past, future, skyline views and the river.

曾经充斥驳船、拖船和有轨车辆的滨水工作区，皇后区的亨特地区在后工业时代逐渐没落。随着最后一艘有轨驳的船的消失，这个曾经壮观的地方沦落为社会蒙羞之地。作为西皇后公园总体规划的一部分，托马斯贝尔斯利协会，连同温特劳布迪多米尼克，把龙门广场州立公园当做可以歌颂其过去和未来，并欣赏海岸线美景之地。

The park is divided into three areas. The Promontory is a great lawn with a natural shoreline edge that takes full advantage of the stunning view of the Manhattan skyline. In North Gantry Plaza, the skyline is framed by restored gantries—gigantic structures that once transferred railcars onto rail barges. Framed by tree-shaded cafés, a fog fountain and game tables, the plaza accommodates 30,000 viewers for the 4th of July fireworks. South Gantry Interpretive Garden is a contemplative space formed by two paths; here, stepping-stone blocks provide the visitor with direct access to the water and look as if they had been abandoned only yesterday. Peninsula Park offers a great lawn promontory with a natural shoreline edge. It is enhanced with willow trees and natural grasses and encourages a wide variety of passive activities, the foremost of which is enjoying the stunning view of the Manhattan skyline.

公园可分为三个区域。海角是一片大草坪，具有天然的海岸线边，可充分享受曼哈顿岛海岸线的壮丽景色。在北龙门广场，地平线被修复的起重机架框住，机架结构巨大，曾经用来把轨道车移到有轨驳船上。广场周围是绿树成荫的咖啡馆、雾状喷泉和玩具桌，广场可容纳欣赏 7 月 4 日焰火汇演的 3 万名观众。南龙门诠释花园是由两条小径围成的思考空间，在此处，垫脚石使游客直接涉水，似乎它们昨天刚被抛弃。半岛公园提供了一个带有天然海岸线边的大草坪岬，柳树和天然的草地更是锦上添花，这里崇尚各种静态活动，其中最主要的就是欣赏曼哈顿岛地平线的壮观景色。

This extraordinary site was blessed with a diverse shoreline and an intact light industrial/blue collar residential neighborhood whose diversity inspired its design. This place serves much broader social purposes by healing a once-divided community and instilling in it a strong sense of neighborhood spirit and pride. That an alliance called Friends of Gantry Plaza State Park was formed by original residents to protect the park is a testament to the power of this place.

此处得天独厚，拥有各式海岸线和完整的轻工业／蓝领住宅区，住宅区的多样性为设计提供了灵感。这个地方愈合分裂的社区，灌输强烈的的互助精神和自豪感，其社会功能更加宽泛。原居民为保护公园组成的、名为龙门广场州立公园之友的联盟，证明了这片地区强大的力量。

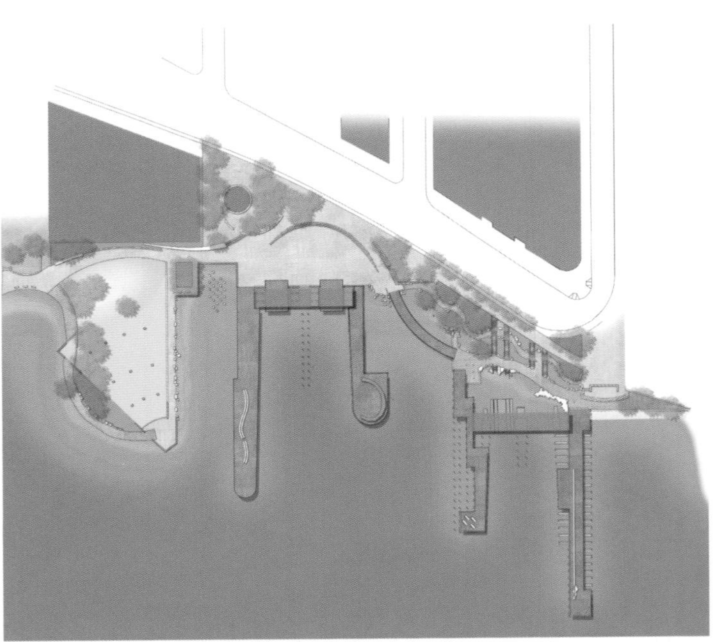

TERRAGRAM

OFFICE PROFILE

Our studio is more a comfortable 'living' space, rather than an office designed for appearances only. The team consists of people with a high degree of tolerance to the moods of the studio patriarch. All individuals maintain their niches of particular skills that make them in-disposable. The real team glue is a compatibility of the values and attitudes to life. We work and (often) play together. Many past colleagues, whilst orbiting elsewhere still keep the umbilical cord – via email or other means. Usually there were 4-5 people in the studio with some others working adjunct on competitions whilst being employed elsewhere.

According to years served, the current team are:
The Elder – Vladimir Sitta, established the company after winning a competition. He is employed for 24 hours a day. Responsible for everything, especially tasks nobody else wants to do. An idealist and cynic in one persona.

The Realist – Robert Faber, who creates documentation miracles which keep the insurance claims at zero. He represents nicety in the studio.

The Dreamer – Anita Madura, the youngest member of the team who believes that changes for the better are embedded in the human psych.

Eva Sitta – the actress, director and accountant that somehow manages to keep us out of jail.

Usual collaborations:
As we do not live in the age of polyglots, we seek and are sought for collaboration by architects, artists and specialists, with complementary skills and values. Our collaborators are mostly based in Australia, however they can be found also in France, Czech Republic and China.

To name a few:
Allen Jack + Cottier Architects (Sydney, Australia)
Chris Elliott Architects (Sydney)
Luigi Rosselli Architects (Sydney)
Oculus Landscape Architects (Sydney)
Reichen et Robert & Associes (Paris, France)
P. Pelcak Architects (Brno, Czech Republic)

Short report:
Whilst the initial impetus for Terragram's existence came from winning a competition (1985), another incentive was to have a working platform that would engage in a critique of the then prevailing pragmatic approach to designed landscape, pursued by most practices in Australia in the mid eighties.

Despite an undeniably increasing scope of tasks solved by landscape architects today, we still consider the narrow definition of landscape architecture quite limiting, especially if forced to operate in pre-determined situations with a scope further reduced by political expediency and their value system. The profession appears hopelessly suspended between sustainability and impotence and cultural sustainability does not seem to be ever entering the formulaic approaches.

We do not consider ourselves a traditional office. The French word 'atelier' better describes the atmosphere and our inclination to experiment, cherish intuitive responses, invite the unexpected and scary, and to technologically innovate. Despite the modest size of the company, our interest range also includes stage design, sculpture, graphics, furniture and educational activities. Our cultural curiosity led us to work in different countries outside Australia. (Bolivia, China, Czech Republic, France, Germany, Indonesia, Iran, Israel, Italy, Japan, Korea, New Caledonia, Singapore, Switzerland, Taiwan, the United States of America and Vietnam).

Chronos

时间花园

LOCATION：Sydney，Australia
项目地点：澳大利亚 悉尼

AREA：700 m^2
面积：700 平方米

COMPLETION DATE：2007
完成时间：2007 年

DESIGNER：Vladimir Sitta，Terragram Pty Ltd
设计师：Vladimir Sitta，Terragram Pty Ltd

ARCHITECT：APA Alex Popov Associates
建筑师：APA Alex Popov Associates

PHOTOGRAPHER：Vladimir Sitta
摄影师：Vladimir Sitta

DESIGN COMPANY：Terragram
设计公司：Terragram

Chronos

This garden sees an exploration of water, from the front of the property to the harbour' s edge at its end. The first floor terrace is an alternative to the now ubiquitous and amorphous green walls provided as an animated back drop—it is a green canvas that will always carry signs of its original structure, clearly outlining areas of dryness and wetness and planted with various rock orchids, bromeliads, ferns, succulents and moss.

从居所的前面到它末端的海港边，这个花园无不体现着对水的开发利用。一楼的平台取代了过去那些在现在看来普通无形的绿色屏障，成为了一个生机勃勃的背景图——平台就是一块绿色的画布，将一直带着它原有的结构图形，清晰地勾划出干燥和湿润的区域，上面种植了各种石兰、凤梨花、蕨类植物、肉质植物和苔藓。

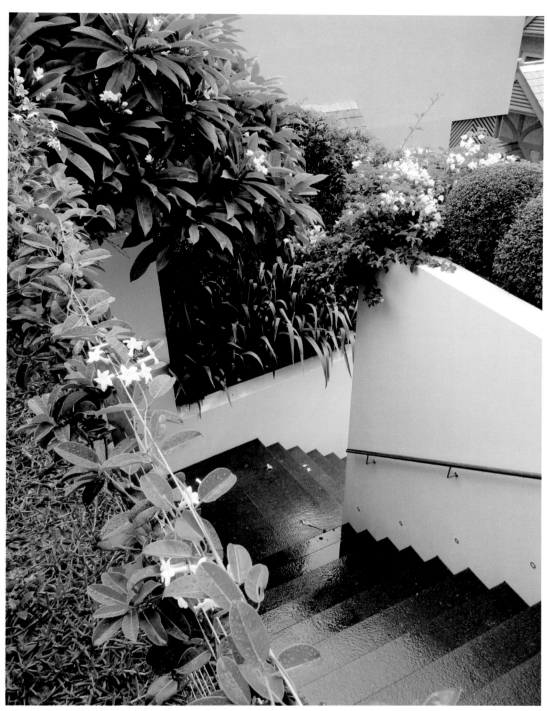

In the main courtyard area, a mechanised tidal pool, with a slightly-submerged brass grate platform stands within a pool of water. The original intention behind this, was the idea that the slightly-submerged platform would become an occasional stage for the owner's grand piano; the garden an incarnation of a "concert hall".
Come night, the garden although plunged into darkness would be alive with music—the illusion of a lone piano almost "levitating" in a platform of water, with soft music resonating. Alternatively, with slight draining of the water level, the brass grate may become an extension of dry surface—another place to pull out some chairs and a table for additional guests. Yet even standing alone, this platform becomes an underwater canvas for the reflection of a bamboo-planted alongside.

在主庭院区，有一个带机械装置的潮水池，里面立着一个稍稍没入水中的黄铜格栅平台。这个设计背后的意图是这个稍微没水的平台偶尔会成为主人华丽钢琴的舞台；花园会化身成为一个音乐厅。
晚上，花园虽然淹没于黑暗中，但音乐让它充满活力——想象一下一架钢琴独自几乎是漂浮在水平台的上方，柔和的音乐在回荡着。或者，把水面稍微排低，黄铜格栅可能变成多出的干燥地方——摆放出几把椅子和一个桌子用于招待额外的客人。但就是独自立在那儿，这个平台也会变成水下的画布，可以倒映旁边的竹子。

Another distinct aspect of this courtyard, is the foreign letters of the Greek word for "time"—"chronos", eating into the stone at the water's edge, only legible in their reflection. This idea of time is here played out, with moments where the word is clearly legible in the still reflection of water and other times where its form is lost in a gentle rippling motion.

这个庭院另一个鲜明的部分是那几个外文字母，组成希腊语中表达时间的那个单词——“chronos”，字母刻在水池边的石头上，只有看它的水中倒影才能辨认出。时间的概念在这儿被演绎了出来，水面平静的时候，字母清晰可认，水面泛起涟漪时，字母就消逝不见了。

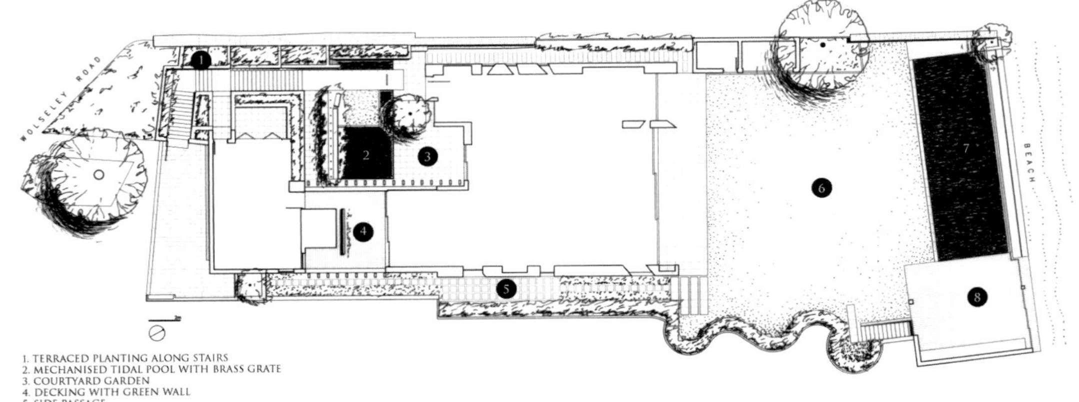

1. TERRACED PLANTING ALONG STAIRS
2. MECHANISED TIDAL POOL WITH BRASS GRATE
3. COURTYARD GARDEN
4. DECKING WITH GREEN WALL
5. SIDE PASSAGE
6. LAWN
7. SWIMMING POOL
8. EXISTING DECK

The substantial backyard area of the property consists of a large lawn, screen planting and swimming pool. The pool forms an edge to the garden, with no visible "boundary" but rather the illusion of the pool merging into the harbour—an "infinity edge".

居所后院重要的区域由一块大草坪、绿色植物屏障和游泳池组成。泳池构成了花园的边际，没有可见的边界，感觉好像泳池与海港合为一体——“无边际”。

Red Garden

红花园

LOCATION：Sydney，Australia
项目地点：澳大利亚 悉尼

AREA：75 m²
面积：75 平方米

DESIGNER：Terragram Pty Ltd
设计师：Terragram Pty Ltd

ARCHITECT：Luigi Rosselli Pty Ltd
建筑师：Luigi Rosselli Pty Ltd

PHOTOGRAPHER：Vladimir Sitta
摄影师：Vladimir Sitta

DESIGN COMPANY：Terragram
设计公司：Terragram

Red Garden

The Red Garden was essentially the result of a collection of "impractical" doodles, an explosion of red sandstone, and an extremely trusting and adventurous client, willing to play the role of a patron, to this experimental garden. In the design phase, clay models were used to help develop the site, which is typical for particularly challenging gardens. There was also a lot of involvement with the site during actual execution, which is a must when gardens do not conform to typical defined geometries. A strip mock-up was constructed on site to verify the various angles and geometries.

红花园实际上是这样产生的：一大批不切合实际的涂鸦，大量暴涨的红沙石以及一个极其信赖别人和爱冒险的客户，他愿意资助这个实验性的花园。在设计阶段，使用了泥塑模型来对该场地进行开发改造，对于特别有挑战性的花园这是典型的做法。在实际施工阶段，还要牵涉很多东西，当花园不符合常规界定的几何尺寸时，这是无法避免的。一个条形的实体模型在工地被建造出来去验证各种角度和几何尺寸。

On entry to the Mosman property, the actual "Red Garden" is initially concealed, and a visitor would first view the swimming pool area. This more subtle space with tiled pool, blue pebbles, small lawn area, and plantings of grasses and bamboo, gives away no clues to the adjacent Red Garden. The cool greys, blues and greens of this space, are a stark contrast to the red and raw dramatics of the Red Garden, which probably heightens the surprise of the visitor upon entry.

一踏入这块位于 Mosman 的地产，红花园最开始是隐蔽的，游客会先看到游泳池区域。这块空间比较精细，水池铺着瓷砖，有蓝色的鹅卵石，小块草坪，栽种了各种草和竹子，没有暴露毗邻的红花园。这一空间里灰、蓝、绿等冷色调与红花园里红色自然态的迷人景致形成鲜明的反差，这可能让一进入红花园的人感到更加惊叹。

The garden contains a number of plant species, many succulents from the client's original collection of haphazard pot plants. The original mature Dracaena draco dragged across Sydney unfortunately died from overwatering after transplantation, and was replaced in 2005 by a mature Pandanus with huge swollen aerial roots. As the garden matures, plants continue to spread and "erupt" from the stone and colonise the garden. Recently the water pool, originally containing fish, has become colonised by duck-weed, forming a "solid" green carpet amidst the red stone—a strangely intriguing reincarnation of the pond.

这个花园里植物品种繁多，许多肉质植物是从客户原先收集的但杂乱栽种的盆栽植物中选出的。最初的那棵横跨悉尼拖回的成年龙血树，由于移植后浇水过多不幸死掉，在 2005 年被一棵成年露兜树取代，它带着膨胀的气生根。随着时间的推移，花园里的植物不断蔓延，从石缝里钻出来，占据整个花园。水池最初是有鱼的，最近被槐叶萍占满了，在红石间形成一块密实的绿色地毯——这是池塘迷人而怪异的重生。

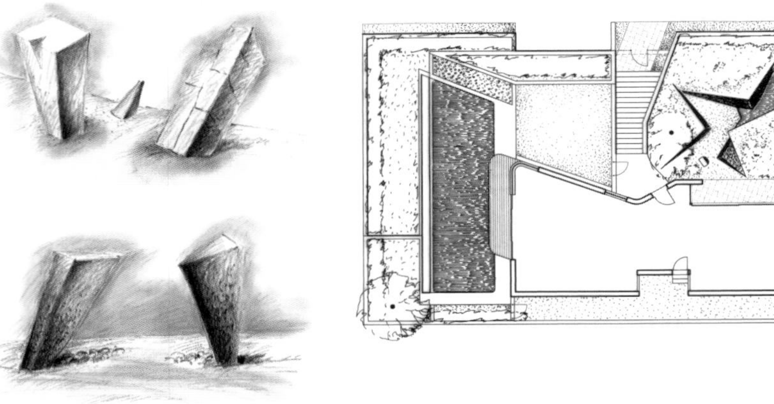

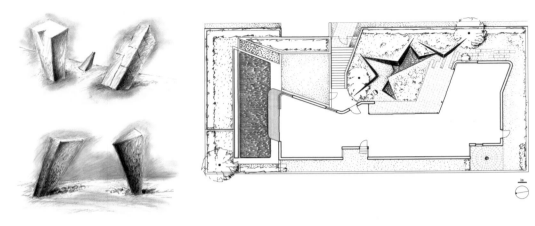

From the interior of the house, the garden is like a painting viewed through the large glass sliding doors. There is a stillness and permanency to the stone, yet at the same time the angled stone is imbued with energy and theatrics.

从屋子里面，透过巨大的玻璃拉门看去，花园像一幅油画。石头是寂静和恒久的，但同时成角度放置的石头蕴含着能量和非凡魅力。

Today the Red Garden continues to thrive, the plants doing extraordinarily well—perhaps sometimes too well. From time to time specialists are invited for maintenance, to ensure the overly rampant plants do not smother the original design intent and geometry of the stone.

今天红花园还是那样欣欣向荣，里面的植物枝繁叶茂——也许有时过了头。偶尔专家会被请来做维护，以确保过度繁茂的植物不会遮盖了原来的设计意图和石头摆放的几何图案。

The small courtyard to the property was originally intended to contain a receding slate cross, however, problems with the availability of gradually increasing stone size, led us to use a spiral form, with a black central hole—almost like Anish Kapoor or Goldsworthy sculpture. Unforeseen by us, it also acted as a bird trap. Lured by water, once inside it became impossible for birds to leave. The current incarnation with planting is a necessary response to stopping the carnage.

该居所的那个小庭院最初打算是放一个向后倾斜的板岩交叉造型，然而，由于石头的体积无法不断增大，我们使用了一个螺旋造型，中间是一个黑洞——很像 Anish Kapoor 或 Goldsworthy 的雕塑。我们没有预料到的是它还起到了捕鸟器的作用。过去，一旦鸟受到水的引诱钻了进去，就不可能离开了。现在将其种上植物对停止这种杀戮是有必要的。

HUGH RYAN LANDSCAPE DESIGN

Hugh Ryan MIDI studied architecture in Dublin in the early 1970s before progressing to landscape design and establishing his own practice, Hugh Ryan Landscape Design in 1977 where he specialises in the design and construction of private gardens.

Hugh's designs have received some measure of recognition both at home and overseas and have been published extensively.

Show gardens have also been a strong interest down the years and Hugh won a silver medal at the Garden Heaven Show at the RDS in 2002 with his garden "La Vie Est Belle", and again in 2009, he won Silver with his controversial garden "Sequoia".

Hugh is a former Chair of the GLDA and up to now was served on the SGD Council. www.hughryan.ie

Altar Ego (Show Garden)

自己的圣坛（展示园）

LOCATION：Emo，Ireland
项目地点：爱尔兰 伊莫

AREA：200 m^2
面积：200 平方米

DESIGNER/ARCHITECT：Hugh Ryan
设计师／建筑师：Hugh Ryan

PHOTOGRAPHER：Hugh Ryan
摄影师：Hugh Ryan

DESIGN COMPANY：Hugh Ryan Landscape Design
设计公司：Hugh Ryan Landscape Design

Altar Ego (Show Garden)

When we come to examine the question of our roots, in particular our Irish roots, it is hard to ignore our pagan and later, Christian roots.

当我们审视根源的问题的时候，特别是我们的爱尔兰根源，就会发现当时的异教和后兴起的基督教根源是很难忽视的。

My installation was an attempt to express some of the symbolism that first found expression in pagan times and was later adopted in the Christian era, and now that we are entering the post Christian, (or at least the post Roman Catholic era), I was striving, (by inviting people's participation), to give voice to a new consciousness.

我的作品旨在表达一些象征意义，它们先是在异教时期出现，到后来基督教时代被接受，现在我们正在进入后基督教时代(或者至少是后罗马天主教时代)，我很努力，(邀请人们来参与)，以表达一种新的意识。

There was a time in history when humans must have felt very small and insignificant on this planet. Today we are still small and relatively insignificant, it's just that there are those amongst us who don't fully appreciate that fact. Despite all our wonderful advances we still don't really know the answer to the big questions—who are we and where do we come from? Why are we here and where are we going?

在人类历史上，肯定有那么一段时间人类感觉自己在这个星球上是很渺小的 。现在我们仍然很小，而且相当微不足道，我们之中有些人并不能完全认清这一事实。即使我们已经取得令人惊奇的进步，也仍然无法解答这些问题——我们是谁？我们从何处来？为什么我们在这里？我们要去哪？

Right around the world, in different places and times our early ancestors strove in their own way to answer these eternal questions. Religion of all kinds grew from these searches, and frequently our yearnings found expression in architecture.

在全世界任何不同的地方，不同的时期，我们的祖先用他们自己的方式奋斗，来回答这些永恒的问题。正是因着这样的探索而兴起各个宗教，并且也经常通过建筑的形式来表现这些渴望。

Ireland is rightly very proud of its ancient monuments such as Newgrange, (although I am yet to be convinced that we understand very much of what is being represented there.)

像新庄园这样的古迹是爱尔兰人的骄傲。(虽然我并不认为人们对这些东西理解多少)

My standing stones and water basin/altar were an attempt to give some expression to these ideas, and I hope that when visitors came to leave their own personal mark they were partaking in a communal act of international dimensions. Modern instant worldwide electronics communications are just a new form of religion for me, because I believe that communication is essential for life and it will bring us along the path to asking even more questions: Who are we and where do we come from? Why are we here and where are we going? Are we going anywhere? Is there anywhere to go?

我设计的立石和盥洗池力图表达这些概念，并且我希望当游客们来到这里留下自己的标记时，他们也正在参与一场国际规模的共同行动。现代即时全球电子通信对我来说是一种新形式的宗教，因我坚信生命离不开沟通，它会让我们思考更多的问题，比如：我们是谁？我们从何处来？为什么我们在这里？我们要去哪？我们要去什么地方吗？有什么地方可去？

Baywatch

海岸救生队

LOCATION：Dublin，Ireland
项目地点：爱尔兰 都柏林

AREA：580 m^2
面积：580 平方米

DESIGNER/ARCHITECT：Hugh Ryan
设计师／建筑师：Hugh Ryan

PHOTOGRAPHER：Hugh Ryan and Ewa Cieslikowska
摄影师：Hugh Ryan and Ewa Cieslikowska

DESIGN COMPANY：Hugh Ryan Landscape Design
设计公司：Hugh Ryan Landscape Design

Baywatch

The whole property runs to some 580m² with a building footprint of 166 m², a front garden of 153 m², a side passage of 16 m² and a back garden of 245 m². Now although my design covered the front, side, back and interior gardens, and that there is a common thread throughout, I have chosen here to concentrate only on the front garden, for I believe that this space stands well enough on its own. With a northerly aspect the garden never the less gets its fair share of sunshine and although only about 7m from the sea it is sitting on a relatively sheltered coastline, however winter storms are not unknown. The public road bounds the property to the N and to the W, and as the front garden is raised some 1.5 m above external ground level and behind a 2 m wall, it is not easily overlooked, (even by the nosiest of passers-by.)

整个房产占地约580平方米，其中建筑面积为166平方米，前花园为153平方米，侧道为16平方米，后花园245平方米。虽然我的设计覆盖了前、侧、后及室内花园，而且彼此呼应，但我在此只想把重点放在在前花园上，因为我认为这部分空间完全自成一体。花园面朝北，光线仍然充足，虽然离海边仅仅大约七米远，冬天有风暴，但它位于相对遮蔽的海岸上。该建筑北侧和西侧是公共车道，前花园比外围地表高出大约1.5米，并有2米高的围墙，不会轻易被外界一览无余。（即使是好奇心特强的路人）

Basically my idea was to invite the seascape into the garden thereby to link the property as closely as possible to its surroundings. To achieve this objective I constructed a split-level water feature, with water in the upper level rising and falling in a tide-like motion. By making water the dominant feature of this garden I was attempting to engage the qualities of open space, movement and reflection; qualities that I believe are fundamental to the seascape. For me and for my clients, the enduring appeal of the seascape is its simplicity was born from its complexity. Tides ebb and flow, with colours and textures that are constantly changing. Very often one only needs to glance at the bay to learn the time of day, the season and perhaps more importantly, what the weather is up to at any given time.

我的根本想法是把海景引入花园，从而尽可能地将整个建筑与周围的环境融为一体。为了实现这一目标，我修建了一个错层的水池，上层的水像潮水一样升起落下。水是花园中最突出的特色，于是我试图将开放空间、流动及反射等元素运用其中，我认为这些都是海景中基本的特征。对于我和我的客户来说，海景持久的魅力在于复杂里透着简单，潮起潮落，色彩斑斓。通常，一个人只需看一眼海湾就能知道一天的时间、季节，更重要的是，可能是任何时间的天气情况。

The water feature is on two levels and a layer of sand and loose pebble covers the floor of both ponds. Water is pumped slowly from the lower to the upper pond and then as if to mimic the tide, the water slowly returns to the lower pond only for the cycle to start once again. The sand in the lower pond is never exposed, but as the water ebbs away from the upper pond it slowly emerges and on dry days the colour changes between wet and dry adding to the effect. I positioned three slabs of black basalt in upper pond. These slabs to me look like sailing boats or even sea creatures, but they are also intended to mirror the shape of Howth Head as it seems to float in the water and when it rains they take on a new dimension and become jet-black mirrors. It seems too obvious to say that reflection is one of the great qualities of water, but it can't be ignored, and here it seems to work well.

水池分两层，水底都是一层沙子和松散的鹅卵石。水慢慢地从下面一层抽到上面一层，再由上面流回下层，循环往复，好像潮来潮往。下层水底的沙子从来是看不见的，但当水流下的时候，上层的沙子就慢慢地显露出来，在晴天，沙子的颜色在干和湿之间变换，十分显眼。我在上层的水池立了三块黑色玄武岩石板，在我看来，它们好像是航行的船只或者是海里的生物，也有爱尔兰霍思岬的外形并且都好像漂浮在水中，雨天的时候，展现出一副新的面孔，像乌黑发亮的镜子。水的反射美本来无须提及，但却不能被忽视，这种美在这里似乎得到了更好的诠释。

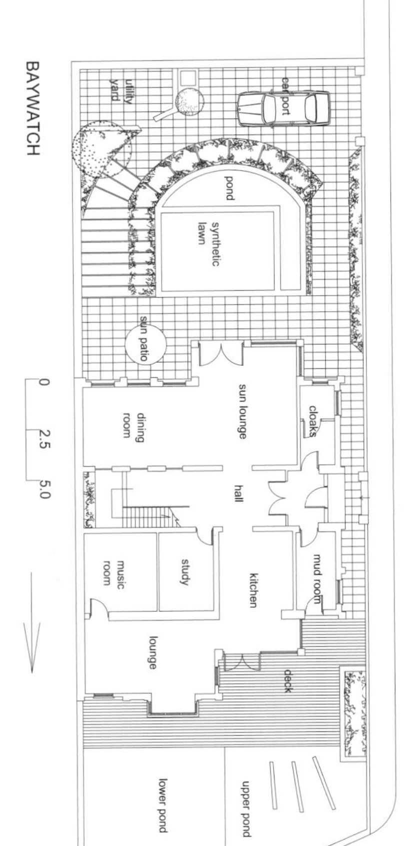

Landfall

着陆之地

LOCATION：Dublin，Ireland
项目地点：爱尔兰 都柏林

AREA：6,000 m^2
面积：6 000 平方米

COMPLETION DATE：2008
完成时间：2008 年

DESIGNER/ARCHITECT：Hugh Ryan
设计师 / 建筑师：Hugh Ryan

PHOTOGRAPHER：Hugh Ryan and Ewa Janczy
摄影师：Hugh Ryan and Ewa Janczy

DESIGN COMPANY：Hugh Ryan Landscape Design
设计公司：Hugh Ryan Landscape Design

Landfall

The broad sweep of Dublin Bay begins on its south side where the ancient fishing harbour of Coliemore is snugly tucked into the area of rocky coastline that is Dalkey Sound. Just off shore it is the craggy outcrop of Dalkey Island that creates this natural place of shelter, and this island with it's own Martello Tower can be seen from anywhere in the bay area.

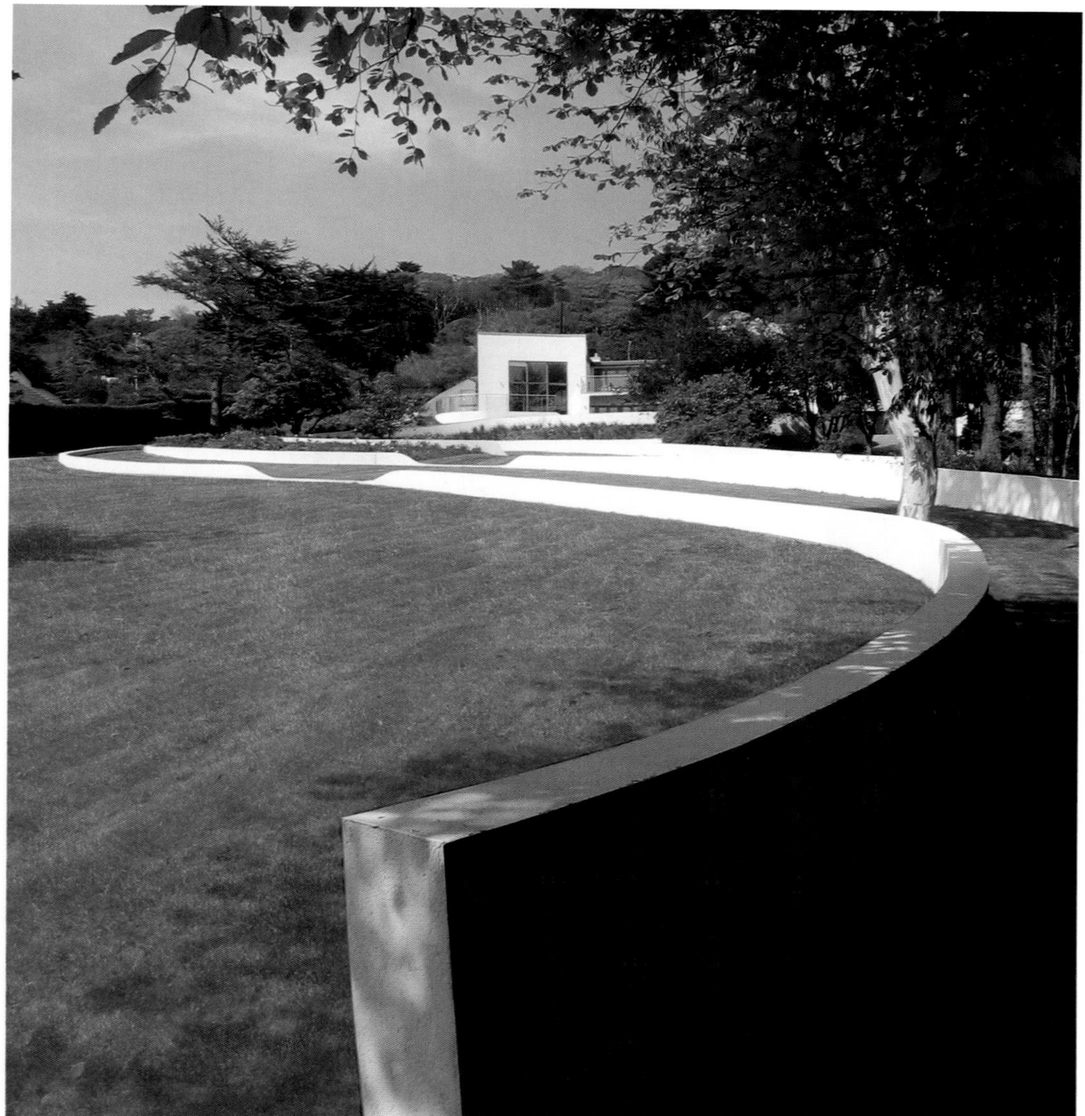

广阔延绵的都柏林湾始于南侧的 Coliemore，它是古老的渔港，紧密的嵌入多基岛的岩石海岸线中。在离海岸的不远处正是多基岛嶙峋的裸露岩石创造出了这一天然避风港，这个岛连同岛上的那个马尔泰洛塔可以在海湾的任何位置看到。

The shoreline of the Bay runs from here, past Dun Laoghaire Harbour and on to Dublin Port itself, before carrying on to form the northern shoreline that culminates in the Hill of Howth. It is on the warm, gentle, south-facing slope of this restful hillside that you will find this project that I call "Landfall", and from here you can gaze back across the water at Dalkey Island and imagine that you are looking down on the Vikings as they made their landfall in Dyflin over 1,000 years ago.

海湾的海岸线从这儿延伸开，经过 Dun Laoghaire 港，一直到都柏林港口，然后在北侧接着延伸，终点是霍斯山。就在这个宁静，温馨、朝南的徐缓山坡上，你会发现这个项目，我称之为“着陆之地”，从这儿你可以隔着海水回望多基岛，并且认为可以窥视着维京人，因为他们 1 000 多年前是在 Dyflin 登陆的。

In the book "*In An Irish Garden*" by Sybil Connolly and Helen Dillon, Olive Gladys Stanley-Clarke described Earlscliffe as a "large ugly house" with a neglected garden overrun with Aubrieta and "a hideous mauve Gladiolus". However, although the Stanley-Clarkes originally had two maids and a gardener, a scarcity of money led them to eventually sell Earlscliffe.

在由 Sybil Connolly 和 Helen Dillon 撰写的《在一个爱尔兰花园中》一书中，Olive Gladys Stanley-Clarke 把 Earlscliffe 描写成一个大而丑陋的房子，带有一个疏于照看的花园，长满了南庭荠和一种难看的淡紫色的剑兰。然而，虽然 Stanley-Clarke 一家人最初有两个女仆和一个园丁，但缺钱导致他们最终卖掉了 Earlscliffe。

Even though she had a dislike for the Earlscliffe house, Olive still loved the Baily area. So they cut a one-and-half-acre corner of the Earlscliffe land off to build themselves a cottage which they named Shiel. She used stones from Earlscliffe to build steps down from the cottage to the lawns of Shiel and planted flowering cherries (dug up from Earlscliffe before they had sold the place).
Kit died in 1983, aged 96. Olive continued to live in Shiel until she sadly passed away on January 26, 1996 at the age of 100.

尽管 Olive 不喜欢 Earlscliffe，但是她仍然爱着这个海湾。于是他们在 Earlscliffe 这块地方开辟出一个 1.5 英亩的角落，为自己搭建了一个小屋，他们叫它“小舍”。她用 Earlscliffe 的石头建筑了台阶，从小舍通向下面的草坪，还种植了开花的樱桃树（在他们卖了 Earlscliffe 之前，在那个地方挖来的）。
Kit 死于 1983 年，终年 96 岁。Olive 继续在小舍生活直到 1996 年 1 月 26 日悲惨地死去，终年 100 岁。

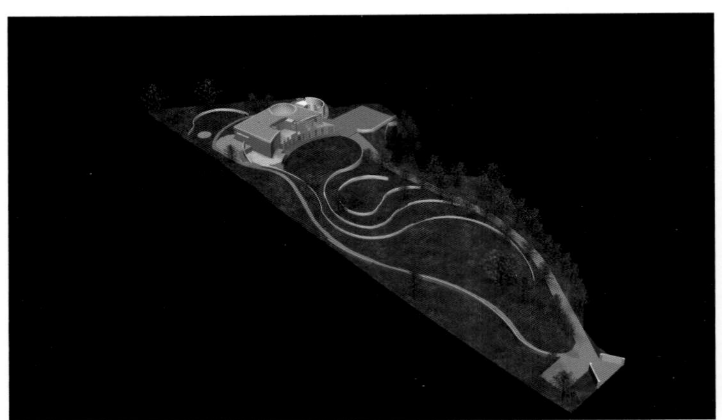
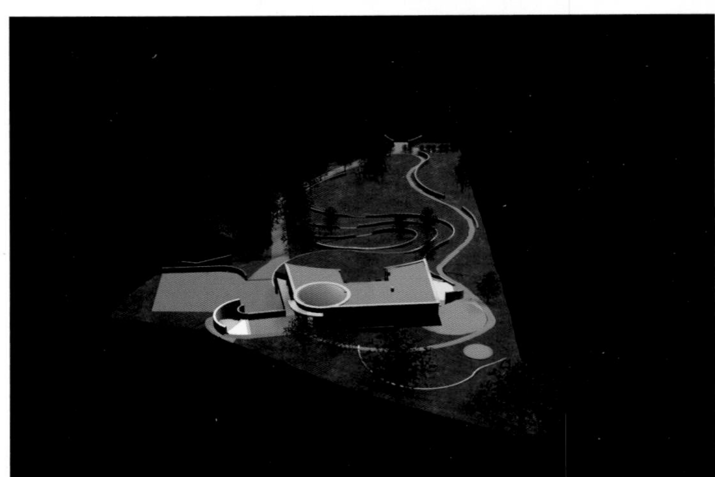

Normandie

诺曼底

LOCATION：Dublin，Ireland
项目地点：爱尔兰 都柏林

AREA：1，400 m^2
面积：1 400 平方米

COMPLETION DATE：2009
完成时间：2009 年

DESIGNER/ARCHITECT：Hugh Ryan
设计师 / 建筑师：Hugh Ryan

PHOTOGRAPHER：Hugh Ryan
摄影师：Hugh Ryan

DESIGN COMPANY：Hugh Ryan Landscape Design
设计公司：Hugh Ryan Landscape Design

Normandie

This garden proved to be a dream project for me, with an enthusiastic and knowledgeable client, a sympathetic and gifted architect and a house to die for. Art Deco as a style has always fascinated me and as a child in the 1950s I was fortunate enough to live quite close to some of Ireland's best examples of Art Deco architecture, and so when I was asked to design a garden for this house in Dublin I didn't have to be asked twice.

这个花园对我来说是一个梦想工程，因为有一位热情而博学的客户，一位体谅人的天才建筑师和一个值得为之尽力的房子。装饰派艺术作为一种风格一直使我着迷，在20世纪50年代，还是孩子的时候，我很幸运，居住在爱尔兰一些最具艺术装饰风格的建筑附近，于是当被邀请为这个位于都柏林的房子设计花园的时候，我爽快地答应了。

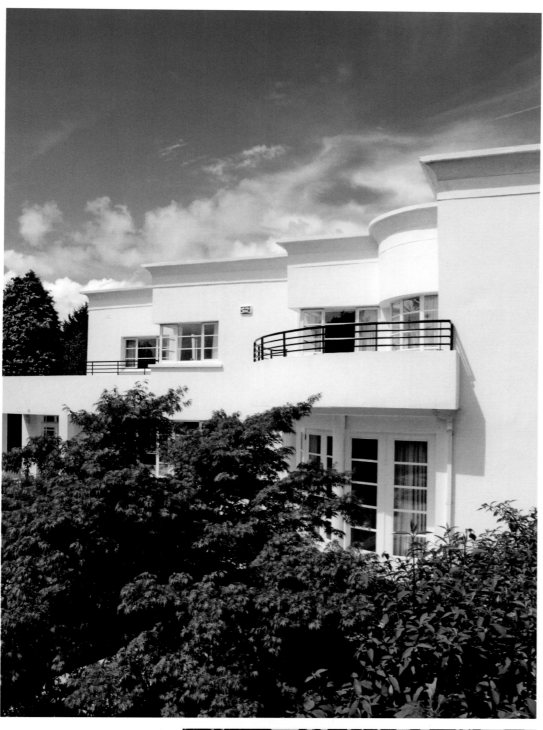

The relationship between the nature of a building along with the space that it occupies, and in turn with the space that surrounds it, is fundamental to my approach to garden design. In this case a large house of obvious character seemed to dominate the site, and my first impression was that it evoked the image of an ocean-going liner, and it was this image that subsequently fueled the inspiration for my design. Later I was to give this project the nickname "Normandie" after the great French Art Deco ocean liner of the 1930s. When I first saw the house it was undergoing a radical renovation and extension and the garden itself was to all intents and purposes a blank canvas, that is to say with the exception of a very fine Siberian spruce Picea omorika in the back garden and a fabulous Cedrus atlantica "Glauca" in the front. I noted that all the rooms in the house had views to the garden, and many on the first floor had balconies. The kitchen occupied an especially important central location in the house and that it provided the principal link between the house and back garden. The client wanted a garden that was neither too traditional nor too modern, a play space for trampoline and basketball, a room to entertain in the back, a lawn and a water feature but not a pond. She also wanted a utility yard to the side and in the front lots of space for cars. She was also keen for me to include some of her favourite plants.

建筑所占的空间和周围空间之间的关系对建筑风格的影响是我花园设计方案中的根本所在。在这个项目里，一个风格鲜明的大房子好像支配着整个场地，我的第一印象是触发人想到远洋轮船的景象，就是这一景象帮助我得到了设计的灵感。后来我势必会把这个项目以20世纪30年代那个伟大的艺术装饰风格的法国远洋轮船取了个别名“诺曼底”。当我第一次看到这个房子的时候，它正在进行彻底的整修和扩建，花园本身实际上就像一张空白的画布，除了后园里一棵非常精美的西伯利亚Picea omorika云杉，和前园里一棵美丽的Glauca北非雪松。我注意到房子所有的房间都能欣赏到花园的景色，二楼的许多房间都有阳台。厨房在房屋里占据了一个特别重要的中心位置，是房子和后园的主要连接点。客户想要一个既不太传统又不太现代的花园，一个玩蹦床和篮球的运动空间，一个位于后面的娱乐房间，一块草坪和一个水景，但不是水池。她还想在侧面要一个干活用的院子，前面要有宽阔的停车空间。她另外强烈要求我加入一些她最喜爱的植物。

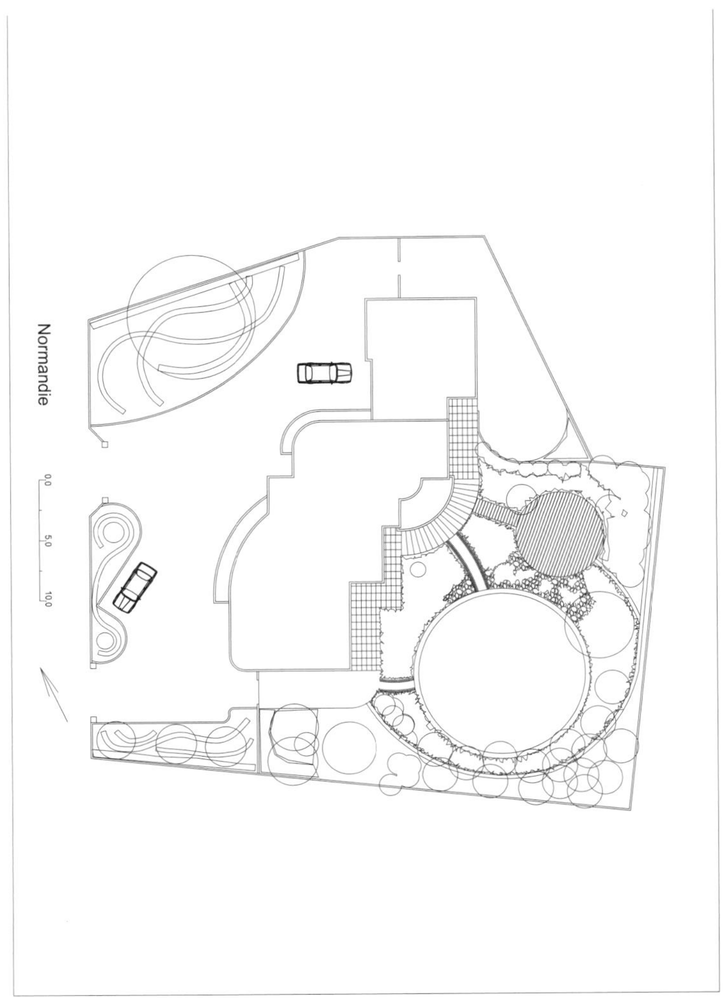
Normandie
0.0
5.0
10.0

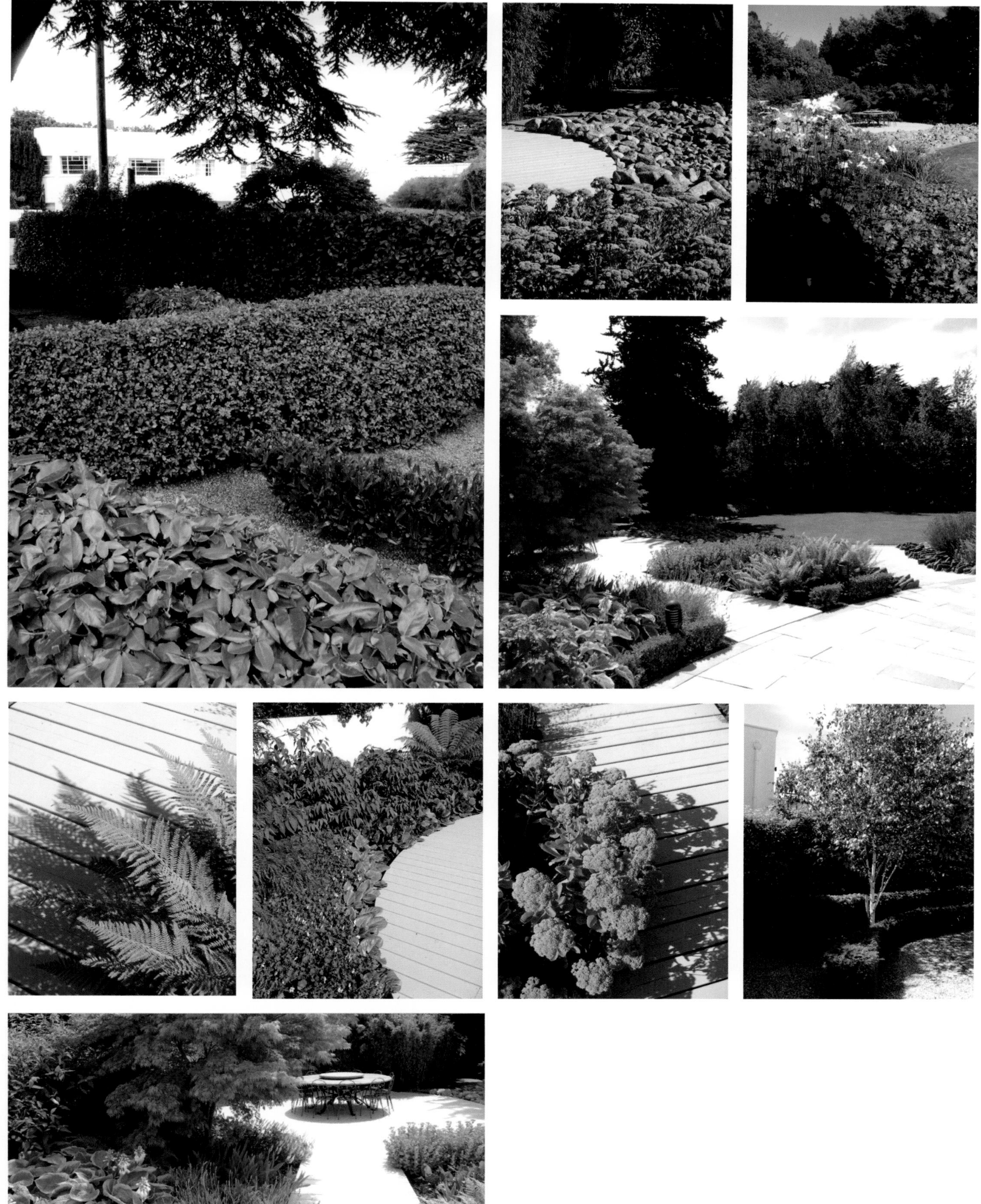

Sequoia (Show Garden)

红杉（展示园）

LOCATION：Dublin，Ireland
项目地点：爱尔兰 都柏林

AREA：80 m^2
面积：80 平方米

COMPLETION DATE：2009
完成时间：2009 年

DESIGNER/ARCHITECT：Hugh Ryan
设计师 / 建筑师：Hugh Ryan

PHOTOGRAPHER：Hugh Ryan and Catherine Ryan
摄影师：Hugh Ryan and Catherine Ryan

DESIGN COMPANY：Hugh Ryan Landscape Design
设计公司：Hugh Ryan Landscape Design

Sequoia (Show Garden)

I like most people value the benefits of tradition in all aspects of our lives, for without it we would have one less "sextant" to guide us on our journey. Gardening has a long and glorious history and it is counted amongst one of the earliest of all human activities.

我像大多数人一样重视传统给我们生活的方方面面带来的益处，因为若没有它，在生命的旅程中指引我们的“六分仪”就是残缺的。园艺有着悠久而光荣的历史，是人类最早的活动之一。

Tradition however is not a static phenomenon and to thrive it needs to be constant on the move. I regret that tradition is oft times confused with orthodoxy or even fundamentalism. This garden was intended to express a number of concepts, but primarily it is hoped that it would hold people's attention long enough to introduce them to an alternative, forward-looking approach to the outdoor spaces which punctuate our urban landscape.

然而，传统不是一个静态的现象，为了繁荣兴旺，它需要不断地发展。我感到遗憾的是传统经常地和正统，甚至是基要主义混淆。这个花园意在表达许多的概念，但是首要的是希望它会长久地吸引人们的注意，在处理那些点缀都市景观的室外空间上，向他们介绍另外一种前瞻性的方式。

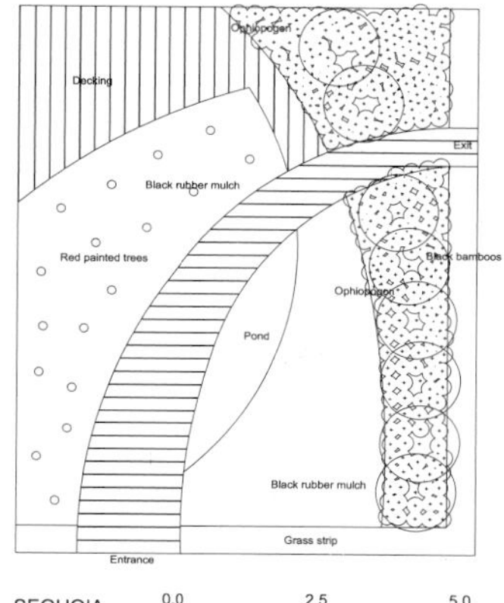

SEQUOIA 0,0 2,5 5,0

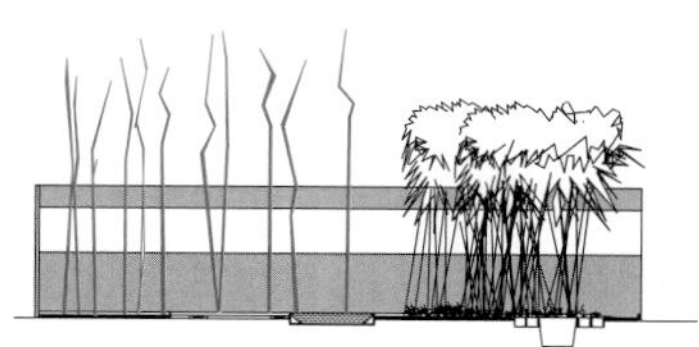

Split Level

错层

LOCATION：South Dublin，Ireland
项目地点：爱尔兰 南都柏林

AREA：680 m^2
面积：680 平方米

DESIGNER/ARCHITECT：Hugh Ryan
设计师／建筑师：Hugh Ryan

PHOTOGRAPHER：Hugh Ryan
摄影师：Hugh Ryan

DESIGN COMPANY：Hugh Ryan Landscape Design
设计公司：Hugh Ryan Landscape Design

Split Level

The original back garden sloped gently away from the back of the house up to where an existing garden studio had been built some years previously. Pushing out an extension to the rear of the house prompted two main considerations.

1. The sloping garden would no longer function comfortably.

2. The garden studio, although is not an eyesore by any means, never the less seemed to loom much larger.

原来的后花园是一段徐缓的斜坡，从房屋的后面开始一直到现在的花园工作室所在的位置，这个工作室建于多年前。对房屋的后面进行扩建主要考虑两点。

1. 缓坡花园起到的功能会大打折扣。

2. 花园工作室虽然绝对不是碍眼的东西，但是它的问题越来越突出。

My solution was to split the garden both horizontally and vertically.

The horizontal split created two levels, one matching floor level in the main house, and the other matching that of the garden studio.

我的解决方案是将花园在水平和垂直方向分割开。

水平分割出两个层面，一个与主房屋的室内地面一致，另外一个与花园工作室的平面一致。

The vertical split, or "split screen" was created by use of a waist-high wall and a bamboo hedge. A central path, which connects the house with the studio, traverses both the horizontal and the vertical splits.

垂直分割或叫做“分割屏幕”是使用一道齐腰高的墙和一道竹子篱笆实现的。中间的一条小路把房屋和工作室相连，横跨了水平和垂直的分割平面。

Raising the middle section of the garden exposed it to view from the neighbours, so I designed and installed a 2.4m-high timber screen to overcome this problem. The screen and the wall were both painted in an orange-coloured latex paint.

抬高花园的中间部分会使它暴露在邻居们的视野范围内，于是我设计并安装了一个2.4 米高的木屏来解决这个问题。墙和屏都喷了橙黄色的乳胶漆。

I used synthetic grass both back and front. In the back a deep-pile lawn with shock pad underneath makes for a low maintenance, all weather and comfortable leisure surface, while in the front, a short-pile, heavy-duty carpet provides an overflow parking space that will always appear green and well kempt.

我在前后侧都使用了人造草。后侧是一片叶子高的草坪，下面带有减震垫，适宜低成本维护，是全天候舒适的休闲场地，前面是短绒耐磨的地垫，提供了看去总是绿色整洁的备用停车空间。

UMBERTO ANDOLFATO

Umberto Andolfato

As a landscape architect, he has a laboratory and studio named L'AB - Lanscape Architecture & Building in Milan. He's been teacher in many schools of design in Milan, and now is Contract Professor of Landscape Architecture at Politecnic of Milan. He received prizes in international and national landscape competitions. Many works were published in a lot of Italian gardens and landscape magazines.

Euroflora 2006

2006 欧洲花展

LOCATION：Genoa，Italy
项目地点：意大利 热那亚

AREA：150 m^2
面积：150 平方米

LANDSCAPE ARCHITECT：Umberto Andolfato
景观设计师：Umberto Andolfato

TEAM：Filippo Fessia，Claudia Mannella，Sara Pallavicini，Katiuscia Panzeri，Francesca Tarditi
设计团队：Filippo Fessia，Claudia Mannella，Sara Pallavicini，Katiuscia Panzeri，Francesca Tarditi

AWARDS：Third Prize in Landscape Section at The Euroflora 2006 Fair in Genova（Italy）
奖项：意大利热那亚 2006 年欧洲花卉交易会景观类三等奖

Euroflora 2006

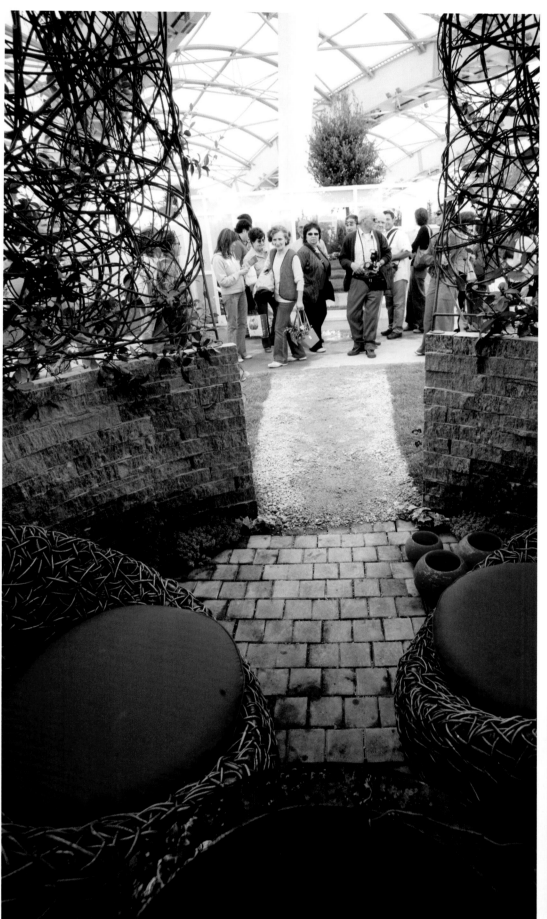

Since the earliest times, the garden has been for man like a small heaven to shelter himself, a "kepos" to find again the inner and outside peace that world refused him.

自古以来，花园对于人类来说就像能够为我们提供庇护所的小天堂。在这里，人类可以再次找回内心的和外在难得的宁静。

That's why we are unceasingly searching for a shelter, for a place to find again that peace of mind which is the basic part of our balance. The "nest garden" wants to mean exactly this.It intends to create a place to protect the nest frame, like a primordial egg which all of us originated from, which surrounds and takes care of us creating that membrane, a filter towards all that surround us and allowing us to sight through its frame the outside world in a nearly detached way.

那就是我们一直在不停地寻找这样一个庇护所，寻找一个能够再次找到心灵安宁和身心平衡的家园的原因。"鸟巢花园"的设计正是想实现这一目标，鸟巢的设计意图是建造一个能够保护鸟巢框架的空间，这是一个初始阶段的蛋形，我们所有的人都源于它，它包围着和照看着我们，它的外膜就是围绕在我们身边的一个过滤器，让我们透过它渴望地观赏外面的世界。

STRUTTURA SALIX
ORGANZA
RETRO
FLAMMA
DAVANTI
SPAZIO ESPOSITIVO
FIAMME LUCI
PALLE DI BOSSO
MACINATO ROSA
MACINATO AZZURRA

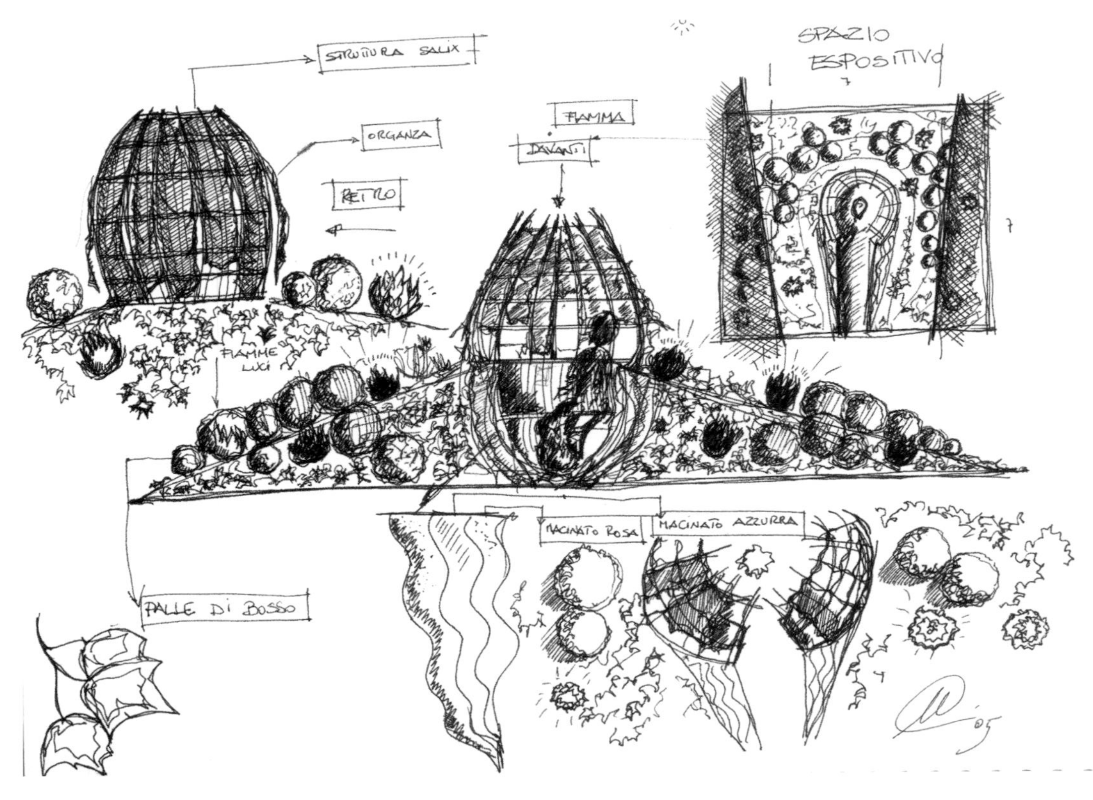

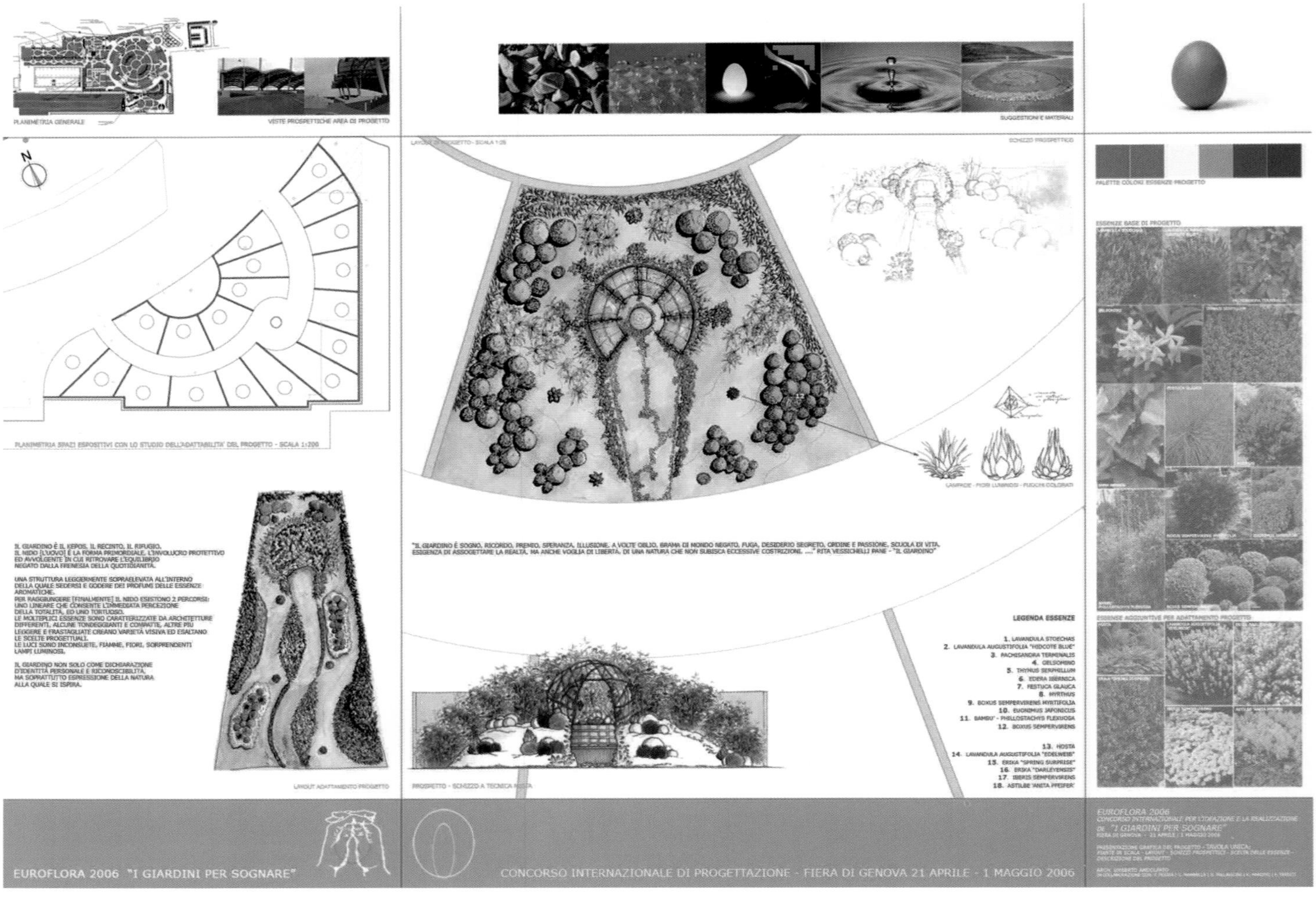

PLANIMETRIA GENERALE
VISTE PROSPETTICHE AREA DI PROGETTO
SUGGESTIONI E MATERIALI
LAYOUT DI PROGETTO - SCALA 1:20
SCHIZZO PROSPETTICO
PALETTE COLORI ESSENZE PROGETTO
ESSENZE BASE DI PROGETTO
ESSENZE AGGIUNTIVE PER ADATTAMENTO PROGETTO
PLANIMETRIA SPAZI ESPOSITIVI CON LO STUDIO DELL'ADATTABILITA' DEL PROGETTO - SCALA 1:200
LAMPADE - FIORI LUMINOSI - FUOCHI COLORATI
IL GIARDINO È IL KEPOS, IL RECINTO, IL RIFUGIO.
IL NIDO [L'UOVO] È LA FORMA PRIMORDIALE, L'INVOLUCRO PROTETTIVO
ED AVVOLGENTE IN CUI RITROVARE L'EQUILIBRIO
NEGATO DALLA FRENESIA DELLA QUOTIDIANITÀ.
UNA STRUTTURA LEGGERMENTE SOPRAELEVATA ALL'INTERNO
DELLA QUALE SEDERSI E GODERE DEI PROFUMI DELLE ESSENZE
AROMATICHE.
PER RAGGIUNGERE [FINALMENTE] IL NIDO ESISTONO 2 PERCORSI:
UNO LINEARE CHE CONSENTE L'IMMEDIATA PERCEZIONE
DELLA TOTALITÀ, ED UNO TORTUOSO.
LE MOLTEPLICI ESSENZE SONO CARATTERIZZATE DA ARCHITETTURE
DIFFERENTI, ALCUNE TONDEGGIANTI E COMPATTE, ALTRE PIÙ
LEGGERE E FRASTAGLIATE CREANO VARIETÀ VISIVA ED ESALTANO
LE SCELTE PROGETTUALI.
LE LUCI SONO INCONSUETE, FIAMME, FIORI, SORPRENDENTI
LAMPI LUMINOSI.
IL GIARDINO NON SOLO COME DICHIARAZIONE
D'IDENTITÀ PERSONALE E RICONOSCIBILITÀ,
MA SOPRATTUTTO ESPRESSIONE DELLA NATURA
ALLA QUALE SI ISPIRA.
"IL GIARDINO È SOGNO, RICORDO, PRESTO, SPERANZA, ILLUSIONE, A VOLTE OBLIO, BRAMA DI MONDO NEGATO, FUGA, DESIDERIO SEGRETO, ORDINE E PASSIONE, SCUOLA DI VITA.
ESIGENZA DI ASSOGETTARE LA REALTÀ, MA ANCHE VOGLIA DI LIBERTÀ, DI UNA NATURA CHE NON SUBISCA ECCESSIVE COSTRIZIONI. ...." RITA VESSICHELLI PANE - "IL GIARDINO"
LEGENDA ESSENZE
1. LAVANDULA STOECHAS
2. LAVANDULA AUGUSTIFOLIA "HIDCOTE BLUE"
3. PACHISANDRA TERMINALIS
4. GELSOMINO
5. THYMUS SERPHILLUM
6. EDERA IBERNICA
7. FESTUCA GLAUCA
8. MYRTHUS
9. BOXUS SEMPERVIRENS MYRTIFOLIA
10. EUONIMUS JAPONICUS
11. BAMBU' - PHILLOSTACHYS FLEXUOSA
12. BOXUS SEMPERVIRENS
13. HOSTA
14. LAVANDULA AUGUSTIFOLIA "EDELWEIS"
15. ERIKA "SPRING SURPRISE"
16. ERIKA "DARLEYENSIS"
17. IBERIS SEMPERVIRENS
18. ASTILBE 'ANITA PFEIFER'
LAYOUT ADATTAMENTO PROGETTO
PROSPETTO - SCHIZZO A TECNICA MISTA
EUROFLORA 2006 "I GIARDINI PER SOGNARE"
CONCORSO INTERNAZIONALE DI PROGETTAZIONE - FIERA DI GENOVA 21 APRILE - 1 MAGGIO 2006
EUROFLORA 2006
CONCORSO INTERNAZIONALE PER L'IDEAZIONE E LA REALIZZAZIONE
DE "I GIARDINI PER SOGNARE"

The elements forming the garden are simple; most of the plants are aromatic essences such as thyme, lavender, rosemary and the box-wood, the absolute plant which "plays" for its shape with the frame of the "nest".
This reference is wanted because large spots of vegetation with box-wood are created as this plant is the hinge of groups of essences.
The area surrounding the "nest" is shaped in a way to create a sloping ground encircling the frame erected in the centre of the space, obtaining in this way a kind of embrace of essences and green and completely winding people seated inside.

组成花园的元素很简单，大部分植被是芳香类植物，像百里香、薰衣草、迷迭香和黄杨木，黄杨木在外形上与鸟巢的结构很相似。
两者之间的联系是必要的，因为大片的植物区都有黄杨木，它在各组芳香植物里起到纽带作用。
鸟巢花园周围的地区以斜坡形式围绕着这个耸立在中心的构架，以这种方式包围着芬芳的花草和其他绿色植物，坐在里面的人们也完全置身于其中。

KEIKAN SEKKEI TOKYO CO., LTD.

Founded in 1986, KEIKAN SEKKEI TOKYO Co., Ltd. (KS) employs a design philosophy seeking the creation of space which harmonizes the natural and built environments, providing for the fundamental human need to be connected to the surrounding world. Applying a process of Total Environmental DEsign Planning & Space Organization (TEDEPSO) we believe successful design celebrates relationships between people, nature, the project site, and the surrounding environment.

KS is dedicated to affecting positive change and humanizing the built environment. A pioneer in the behavioral multi-disciplinary approach to environmental planning and design, KS ensures that environmental qualities are enhanced within the framework of financial, technical and community considerations. Special care is given to assuring that program elements and physical components of a project are mutually supportive, furthering established public policy and management goals, and ensuring the long-term success of every project.

From our studio in Tokyo KS serves a variety of public and private sector clients, successfully realizing projects locally and internationally. Our work includes various scales and project types including ecological & regional systems, master planning, resort planning & design, campuses & corporate headquarters, urban planning & design, civic parks, plazas, and intimate gardens. Our expertise and enthusiasm with the collaborative design process encourage owners, stakeholders, and communities to be active in all phases of the project, seeking consensus and a design solution capable of satisfying diverse needs.

Composed of designers professionally trained in planning, landscape architecture, urban design, engineering, and horticulture, with many formally trained in more than one discipline, KS is able to design and execute projects with care and craft maintaining a focus on personal service and high quality design that is both sensitive and responsive to its location. From reenergizing cities and communities to managing complete landscapes and creating practical and inspiring public spaces, all of our work is defined by a commitment to innovation, design excellence, social inclusion, stewardship of the environment, and provids value to clients and the public.

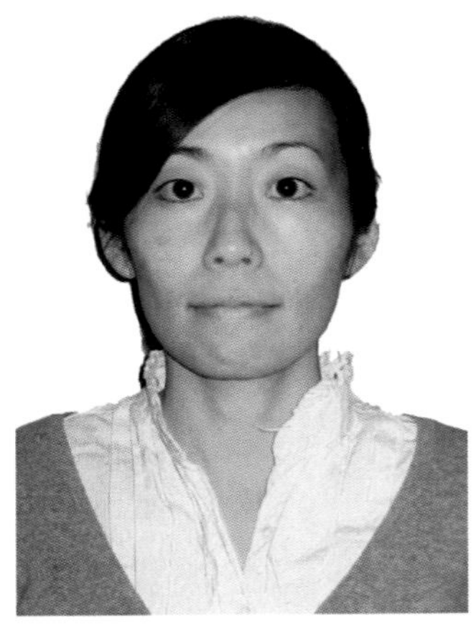

Aya MIYAKODA

Landscape Architect
KEIKAN SEKKEI TOKYO Co., Ltd.

Career:
2011 to Present Landscape Architect: Keikan Sekkei Tokyo Co., Ltd.
2010 – 2011 Landscape Architect: Obayashi Corporation
2005 – 2010 Designer: LandDesign, Inc. Charlotte, North Carolina
Summer 2003, 1999 & 1998 Student Intern: Keikan Sekkei Co. Ltd., Tokyo, Japan
Spring 1998 & 1996 Student Intern: Keikan Sekkei Co. Ltd., Tokyo, Japan
Summer 2000 Student Intern: Ron Herman Landscape Architects, San Leandro, CA
Spring 2000 & 1999 Student Intern: Ichiura Planning and Housing Consultants Co.,Ltd., Osaka, Japan
Spring 2000 & Summer 1997 Student Intern: Noriaki Okabe Architecture Network, Tokyo, Japan

Education:
2004 Master of Landscape Architecture Louisiana State University, Baton Rouge, Louisiana
2001 Bachelor of Design, Environmental Design and Architecture Kobe Design University(KDU), Kobe, Japan

Selected Honors:
2008 Winning Entry—Xing-Gang Landscape Boulevard Streetscape Competition, Suzhou Industrial Park Administrative Committee
2008 Third Place—Sustainable Design Competition, LandDesign
2004 & 2002 Student Merit Award—American Society of Landscape Architects Louisiana Chapter
2004 Design Award and Teaching Award—School of Landscape Architecture at Louisiana State University
2002 & 2003 Helen A. Reich Memorial Award—School of Landscape Architecture at Louisiana State University
2001 Selected to exhibit graduation project for the design university association of western Japan
2000 First Prize—Environmental Design at Kobe Design University

Professional Experience:
Transit-Oriented Development
Hadley | Charlotte, North Carolina (24.6 ac)

Resort / Recreation Development
Zhuhai Qinglv Road Corridor Planning and Design Vision | Zhuhai, China (108 ha)

Urban Development
Brooklyn Village | Charlotte, North Carolina (16 ac)
Brooklyn Park (2nd Ward Park) | Charlotte, North Carolina (3.5 ac)
Suzhou Xing-Gang Streetscape | Suzhou, China (30 sqkm lakefront)

Housing Development
Shenyang Woody Ridge | Shenyang, China (12.4 ha)
"Riverie" Minatocho | Kawasaki City, Kanagawa Prefecture (3.6 ha)

C. Brian BENNETT

Landscape Architect
KEIKAN SEKKEI TOKYO Co., Ltd.

Career:
2010 to Present Landscape Architect: Keikan Sekkei Tokyo Co., Ltd.
2009 – 2010 Principal: C. Brian Bennett Design Studio, Charlotte, North Carolina
2004 – 2008 Designer: LandDesign, Inc. Charlotte, North Carolina
2002 – 2004 Landscape Architect in Training: JFDS, Baton Rouge, Louisiana
2000 – 2001 Director of Landscape Design: Paul Banner Nursery & Landscape, Charlotte, North Carolina

Education:
2004 Master of Landscape Architecture Louisiana State University, Baton Rouge, Louisiana
2000 Bachelor of Science, Horticulture North Carolina State University, Raleigh, North Carolina

Selected Honors:
2002 & 2003 Helen A. Reich Memorial Scholarship Award
2002 Landscape Architecture Endowment Award

Professional Experience:
New Town Development
Victoria Lake Town Centre | New Kampala Uganda (250 ha)

Resort / Recreation Development
Curacasbaai | Curacao, Antilles Netherlands (108ha)

Retail / Lifestyle Centre
The Bridges at Mint Hill | Mint Hill, North Carolina (215 ac.)
Blakeney | Charlotte, North Carolina (300 ac)

Office Development
LNR South Park | Charlotte, North Carolina (25 ac)

Parks & Recreation
Catawba River Pedestrian Bridge | Charlotte, North Carolina
Romare Bearden Park | Charlotte, North Carolina (5.4 ac)

Housing Development
Wei Gou Villas | Beijing, China (22.4 ha)
Millbridge | Waxhaw, North Carolina (928 ac)
Cureton | Waxhaw, North Carolina (700 ac)
River Trail | Charlotte, North Carolina (500 ac)
Tranquil Court | Charlotte, North Carolina (2.5 ac)

Urban Planning
Dubailand Plots 26 & 27 | Dubai, United Arab Emirates (357 ha)
Suzhou Xing-Gang Streetscape | Suzhou, China (30 sqkm lakefront)

Residential Design
Wutschel Residence | Atlanta, Georgia (10 ac)
Norris Residence | Charlotte, North Carolina (1 ac)

Hiroshi WATANABE

Director
KEIKAN SEKKEI TOKYO Co., Ltd.

Career:
1998 to Present Director: Keikan Sekkei Tokyo Co., Ltd.
1996 – 1998 Senior Landscape Architect: Keikan Sekkei Co., Ltd. (Tokyo office)
1990 – 1996 Landscape Architect: Keikan Sekkei (S) Pte. Ltd. (Singapore)
1988 – 1990 Landscape Architect: Keikan Sekkei Co., Ltd. (Tokyo office)
1985 – 1988 Landscape Architect: Zoen Design, Singapore

Select Honors:
2006 2nd Price—Gardens by the Bay, Singapore

Education:
1985 Bachelor of Agriculture Osaka Prefecture University College of Agriculture (Landscape Architecture)

Professional Experience:
Landscape Master Planning
Cyberjaya Landscape Master Plan | Malaysia

Resort Recreation Development
Singapore Zoological Gardens | Singapore

Parks & Recreation
Granpark Plaza | Tokyo, Japan
Green Hills Tsuyama | Okayama, Japan
Hotarumibashi Park | Yamanashi, Japan
Higashi Shinagawa Kaijo Park | Tokyo, Japan
Shinagawa Chuo Park | Tokyo, Japan

Housing Development
Chai Chee Estate Housing Renovations | Singapore
Park City Shin-Kawasaki Housing Project | Kanagawa, Japan
Ideal City Housing Project | Shenyang, China
Wei Gou Villas Housing Project | Beijing, China

Campus Planning
Chinese High School | Singapore

Hotels
Sedona Hotel | Manado, Indonesia

Kinoto MIYAKODA

Design Director
KEIKAN SEKKEI TOKYO Co., Ltd. + TBG Partners Tokyo Office

Career:

2010 to Present Design Director: Keikan Sekkei Tokyo Co., Ltd. + TBG Partners, Tokyo Office: Landscape Architecture and Planning, Tokyo, Japan
2006 – 2010 Associate: TBG Partners: Landscape Architecture and Planning, Houston, Texas
2005 – 2006 Designer: Asakura Robinson Company: Landscape Architecture, Urban Design and Community Planning, Houston, Texas
Fall 2000 Student Intern: GGLO, LLC: Architecture, Interior Design, Master Planning and Landscape Design, Seattle, Washington
Summer 2000 Student Intern: Kajima Corporation, Tokyo, Japan
Spring 1998 & 1996 Student Intern: Keikan Sekkei Co. Ltd., Tokyo, Japan
Summer 1997 – 1999 Student Intern: Keikan Sekkei Co. Ltd., Tokyo, Japan

Education:

2005 Master of Landscape Architecture Louisiana State University, Baton Rouge, Louisiana
2001 Bachelor of Architecture Degree Miyagi University, Sendai, Japan

Selected Honors:
2010 "Best Overall" Award – Rice Design Alliance Competition
2009 Greater Houston Builders Association Houston's Best 2009
2008 Houston Business Journal Landmark Award 2008 for Telfair Community Development
2006 2nd Place – Gardens by the Bay International Competition in Singapore
2005 Thesis Award – Robert Reich Graduate School of Landscape Architecture, Louisiana State University
2004 Honor Award for Academic Achievement – ASLA Louisiana Chapter
2004 Computer Technology Award – Robert Reich Graduate School of Landscape Architecture, Louisiana State University
Feb. 2004 Selection for exhibition in the first-ever CoAD Student Design Show
2002 & 2003 Helen A. Reich Memorial Scholarship Award

Professional Experience:

Resort & Hospitality
Wind Creek Atmore Casino Resort | Atmore, Alabama, USA
Sycuan Casino Hotel | San Diego, California, USA
Pauma Hotel & Casino | Pauma Valley, California, USA
Yoron Island Hotel | Okinawa, Japan

Urban Development
Methodist Hospital Dunn Healing Garden | Houston, Texas, USA
Julia Ideson Library | Houston, Texas, USA

Community Development
Bridgeland | Houston, Texas, USA
Telfair | Houston, Texas, USA
Eastern Regional Park | League City, Texas, USA
Wei Gou Villa, Condominium Project | Beijing, China

Tooru MIYAKODA

CEO
Fellow American Society of Landscape Architects (FASLA)
Principal Landscape Architect / Planner
KEIKAN SEKKEI TOKYO Co., Ltd.

Career:

1998 to Present CEO: Keikan Sekkei Tokyo Co., Ltd.
1986 – 1998 Director / Manager: Tokyo Office Keikan Sekkei Co., Ltd.
1974 – 1986 Kajima Corporation
1973 Full time staff: Zion & Breen Associates Inc.
1971 – 1972 Sasaki Walker Associates Inc. (6 months)
1971 – 1972 Sasaki Demay Dawson Associates Inc (8 months)
1970 – 1971 Part-time staff: Eckbo Dean Austin & Williams Office (3 months)
1967 – 1970 Kajima Corporation

Education:

1967 Master of Landscape Architecture Graduate School, University of Osaka Prefecture
1965 Bachelor of Landscape Architecture University of Osaka Prefecture, Osaka Japan
1971 – 1972 Visiting Scholar Graduate School of Environmental Design University of California, Berkeley
1972 – 1973 Special Student Course Graduate School of Design, Harvard University

Selected Honors :
1984 JILA Award (design) – Osaka Gakuin University
1986 Kawasaki City Landscape Award – Park City Shinkawasaki
1989 Japan Award of Parks and Open Space Association – Ichizawa Park, Saitama
1992 Midori no Toshi Award of Parks and Open Space Association – Granpark Project
1997 Green City Award of Yomiuri News Paper – Granpark Plaza
2001 Sainokuni Saitama Landscape Award – Matsubara Danchi Station Nishi-guchi Park
2001 CLA Award—Hotarumibashi Park
2002 Tokyo Parks Association Award – Higashi-Shinagawa Kaijo Park
2002 ASLA (design) Honor Award – Hotarumibashi Park
2004 CLA Award – Matsubara Danchi Station Nishi-guchi Park
2005 CLA Award – Genecity Housing Project
2006 Merit Award—228 National Park International Design Competition
2006 2nd Prize—Gardens by the Bay International Competition (Singapore)
2007 Fellow American Society of Landscape Architects (FASLA)
2008 Kitamura Award (Japanese Olmsted Award)

Green Hills TSUYAMA

津山绿山公园

LOCATION: Tsuyama, Japan
项目地点：日本 津山

AREA: 290,000 m^2
面积：290 000 平方米

COMPLETION DATE: 2010
完成时间：2010 年

PHOTOGRAPHER: Keikan Sekkei Tokyo Co., Ltd.
摄影师：Keikan Sekkei Tokyo Co., Ltd.

AWARD: Aging Award (Design), Japanese Institute of Landscape Architects
奖项：Aging 奖（设计）日本景观建筑师学会

LEAD DESIGNER: Tooru Miyakoda, FASLA
首席设计师：Tooru Miyakoda, FASLA

DESIGN COMPANY: Keikan Sekkei Tokyo Co., Ltd.
设计公司：Keikan Sekkei Tokyo Co., Ltd.

Green Hills TSUYAMA

Completed almost 15 years ago, Green Hills TSUYAMA displays a socially and environmentally sustainable approach to design and planning prior to these concepts becoming an integral component of every project.

津山绿山公园项目大约 15 年前完成，向我们展示了一个不仅在社会方面同时在环境方面体现可持续发展的设计和规划方案，这些概念将成为津山每一个项目里密不可分的组成部分。

Constructed as a 21st century Lifestyle Park, with careful planning and preservation of original landform, hydrology, and vegetation, the park has evolved with the city and surrounding wild mountain landscape emerging as a single beautiful organism.

津山绿山公园经过精心规划和对原始地貌、水文和植被的保护，成为一个具有 21 世纪生活方式的公园，并在不断发展着，与该市和周围的荒山景观正融合为一个美丽的有机体。

The 29 ha Green Hills TSUYAMA Public Park is located in the heart of Tsuyama, Japan serving a local population of over 110,000. The city, dating back over 400 years, is primarily located north of the Yoshii River in a natural basin created by mountain ranges surrounding the city on all sides. As a result, the city has grown in a linear fashion northward through the valley from the river.

日本津山绿山公园占地 29 公顷，坐落于日本津山的心脏位置，为当地 11 万多的居民提供服务。这座城市可以追溯到 400 多年前，主要坐落于吉井河的北部，地处一个天然盆地内，盆地是由城市四边的山脊环绕而成。其结果是，该城市沿着河流穿过了的山谷，以线形方式向北发展，

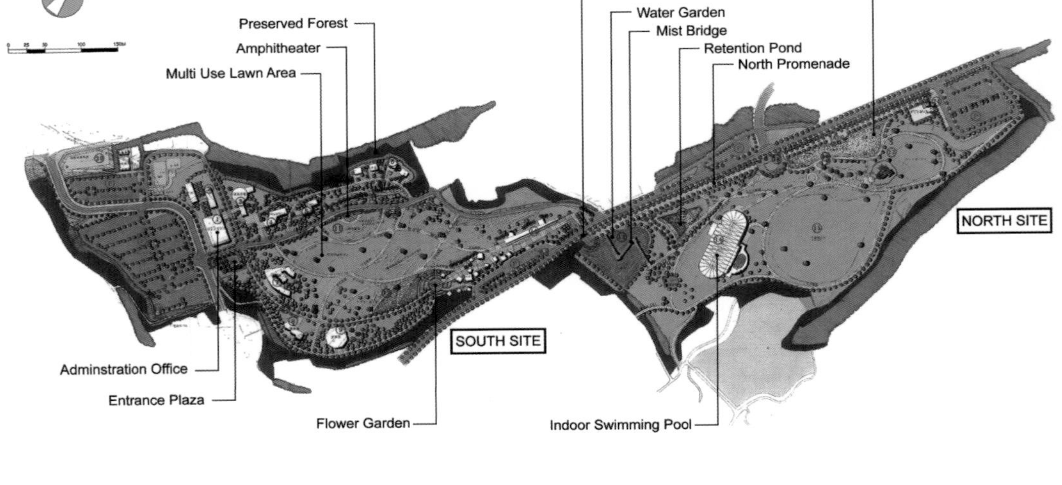

Due to an abundance of natural wilderness, recreational areas surrounding the city, or large public parks capable of accommodating programmed events were long neglected in the city. As the population of Tsuyama grows, the need for such a space increased while the availability of suitable land decreased.

由于大面积的自然荒地，城市周围的休闲区，也就是有能力承办正式活动的大公园长期受到忽视。随着津山人口逐年增加，对于这种空间的需求增加了，而合适的可用土地却减少了。

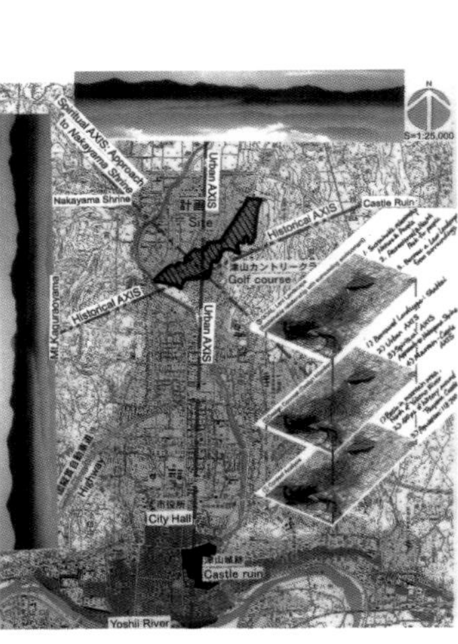

When a large parcel of government land became available for redevelopment, officials recognized the opportunity to provide this much-needed regional park capable of accommodating large cultural events and festivals and providing recreational opportunities not available in the surrounding wilderness.

The site formerly served as the Institute of Dairy Farming of Okayama Prefecture. The existing landscape was pastoral, characterized by large expanses of rolling grassed fields spotted with small stands of mature trees. The elevation at the high points of the site provided ideal views of the city below and mountain ranges surrounding the city in the distance.

当一大块政府土地可再开发时，官员意识到提供这个区域公园十分必要，它能够为大型文化和节日活动提供场所，带来以前在周围荒野地区不能得到的娱乐机会。

该地点以前是冈山县乳液农业研究所，现存的田园景观，有连绵的宽阔草地，上面点缀着小棵的成年树木。从这里的制高点俯瞰下去可以看到下面的城市中最理想的风景以及远处环绕城市的山脉的风光。

Hotarumibashi Park

Hotarumibashi 公园

LOCATION：Minami Alps，Japan
项目地点：日本 南阿尔卑斯

AREA：7，200 m^2
面积：7 200 平方米

COMPLETION DATE：2010
完成时间：2010 年

PHOTOGRAPHER：Keikan Sekkei Tokyo Co.，Ltd.
摄影师：Keikan Sekkei Tokyo Co.，Ltd.

AWARD：Honor Award（General Design），American Society of Landscape Architects
奖项：荣誉奖（设计） 美国景观设计师协会

LEAD DESIGNER：Tooru Miyakoda，FASLA
首席设计师：Tooru Miyakoda，FASLA

DESIGN COMPANY：Keikan Sekkei Tokyo Co.，Ltd.
设计公司：Keikan Sekkei Tokyo Co.，Ltd.

Hotarumibashi Park

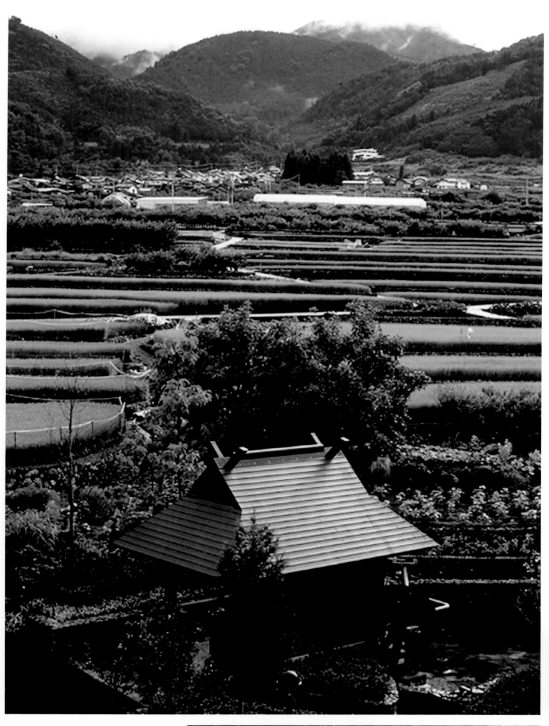

The primary objectives for the park were creating harmony with the surrounding environment, and providing a means for the community to interact and rediscover the rural landscape of the terraced rice fields, weaving together agriculture, community, and ecology. Site elements consist of the Mount Fuji viewing plaza, terraced plaza, community centre, recreated soba mill, educational facilities for schoolchildren, play areas, family picnic areas, parking, a path system, new planting, two eco-ponds, and the reconstruction of the lost habitat of the fireflies that gave the park its name (Hotarumibashi is translated as "firefly viewing bridge").

公园的主要目标是与周围环境建立和谐关系，并为该居民区的社交活动提供便利，再现水稻梯田的乡村景观，把农业、社区和生态结合在一起。内容包括富士山观景广场，梯田广场，社区中心，重建的荞麦磨坊，为学生新建的教育设施，游玩区，家庭野餐区，停车场，道路系统，新栽的植物，两个生态水塘，并为萤火虫重建栖息地，公园因此得名（Hotarumibashi 翻译为“萤火虫观景桥”）。

The heart of the design is a two-meter wide terraced plaza, which mimics the terraced agrarian landscape and is constructed of the same stone used in the rice fields, taking visitors from the roadside down to the edge of the river. There are six major sections in three switchbacks, with an observation plaza at the midpoint of each switchback. Each section functions as a gallery utilizing shakkei, or "borrowed landscape" to feature the rice terraces and hills adjacent to the park, and Mount Kushigata in the background from different angles and elevations.

设计的核心是一个 2 米宽的梯形广场，模仿了梯田农业景观，由水稻梯田使用的那种石头修建而成，可以让游客从路边下到河边。该广场由三个之字形组成，有六个主要部分，每个之字的中点拐点处是一个观景广场，每个部分就像一个画廊一样，利用借景功能从不同角度和高度，在画廊里欣赏水稻梯田、毗邻的山丘以及 Kushigata 山峰等景致。

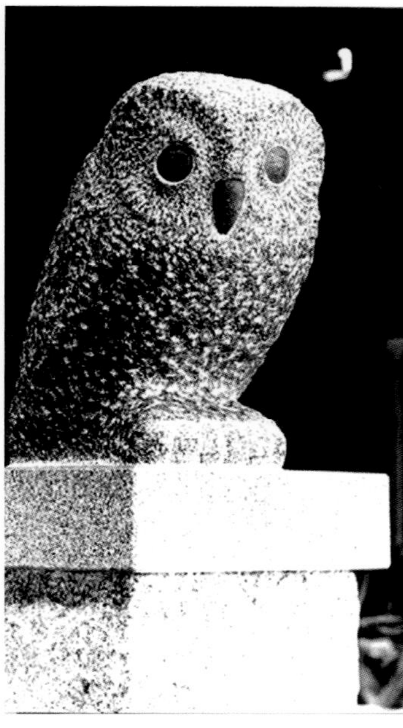

The result is a series of images that move from a panoramic sweep of the surrounding countryside to an intimate feeling of being immersed within the landscape. A constructed stream beginning at the base of the ramped plaza plays an important part in the ecology of the park. When the road project relocated the river into its current cement and stone channel, the habitat for the fireflies was destroyed. By diverting river water into an artificial streambed constructed using the same materials and plantings that had originally formed the natural bed of the river, the firefly habitat was restored. This stream leads visitors through the park, eventually feeding into the iris-pond and interactive-pond which appear as part of the hydrologic network constructed to irrigate the rice fields down slope. Located at the lowest point of the site, these eco-ponds allow children to play in the water and explore the firefly habitat firsthand, while also retaining stormwater runoff, and improving water quality.

其结果是你会看到一个系列的图景，从一个农村环境的全景扫描到融入自然乡村景观的亲密感觉。在斜坡广场底部流出一条人工建筑的小河，它在这个生态公园中发挥了重要作用。当道路工程把河水位置移到现在的水泥和石头通道中时，萤火虫的栖息地被破坏了。采用与当初天然河床相同的材料和植物建造了一条人工河，将河水导入其中，使萤火虫栖息地恢复。这条河流引领游客穿过公园，最终汇入菖蒲池和交互池，两个水池是为灌溉坡下稻田而建的水利网络的一部分。这些生态塘坐落在公园的最低处，可以让孩子在水里嬉戏，直接探索萤火虫栖息地，同时也保留了径流的雨水，并改善水质。

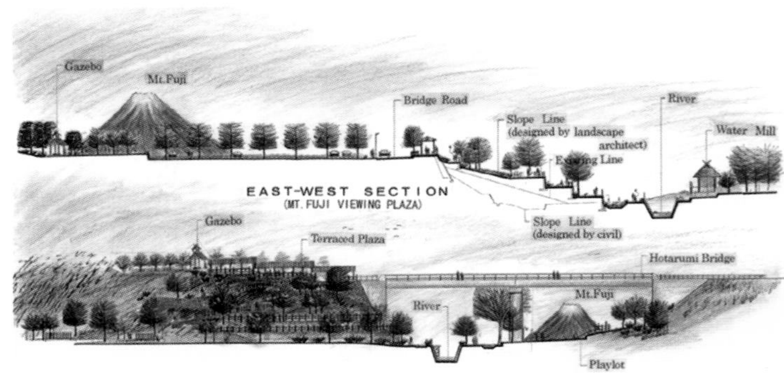

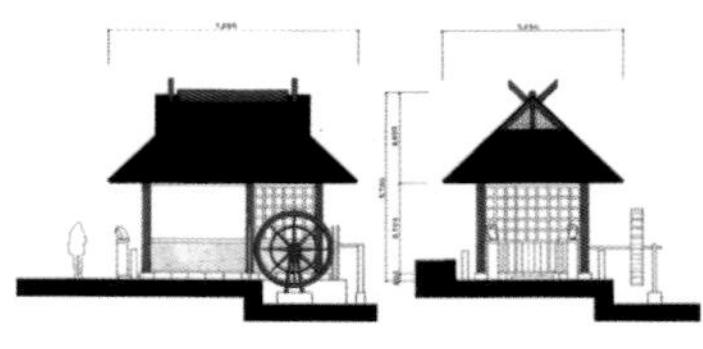

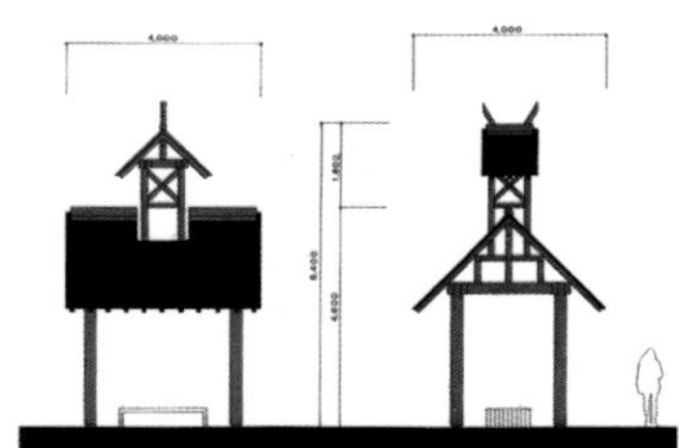

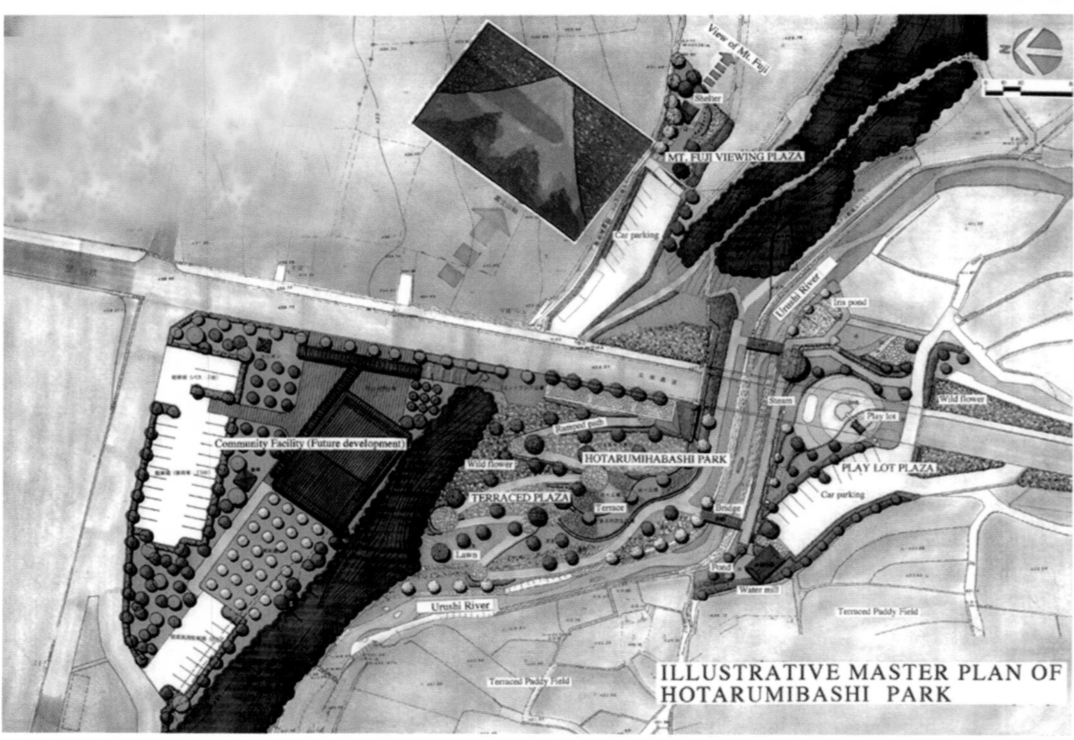

ILLUSTRATIVE MASTER PLAN OF HOTARUMIBASHI PARK

On the hilltop above these eco—ponds lies the Mount Fuji Viewing Plaza. Here a wooden structure designed to represent a samurai helmet is rotated away from the orientation of the entry path and aligned to provide a direct view to Mount Fuji by a technique known as ikedori, literally the "capturing alive" of outside elements and incorporating them into the design. By working with the local flora and fauna, local materials, the surrounding scenery, and the cultural and historical symbols of the region and nation, Keikan Sekkei transformed scraps of land into a park that has become the pride of the local community

这些生态塘上方的山顶上是富士山观景广场。在这里，有一个设计成代表武士头盔的木制结构，它扭身朝远离入口路径的方向排列，用一种称为 ikedori 的技术提供富士山的直接风景，这一技术字面上叫"活捉"外部元素并把它们纳入其设计中。通过研究当地的植物和动物、当地材料、周围的风景及这个地区和国家的历史文化象征，Keikan Sekkei 景观公司把一块块小小的土地变成了一个美丽的公园，它已成为当地社区的骄傲。

Fukuoka—Zeki Sakura Park

福冈堰樱花公园

LOCATION：Ibaraki Prefecture，Japan
项目地点：日本 茨城县

AREA：27,000 m^2
面积：27 000 平方米

COMPLETION DATE：2010
完成时间：2010 年

LEAD DESIGNER：Tooru Miyakoda，FASLA
首席设计师：Tooru Miyakoda，FASLA

DESIGN COMPANY：Keikan Sekkei Tokyo Co.，Ltd.
设计公司：Keikan Sekkei Tokyo Co.，Ltd.

Fukuoka—Zeki Sakura Park

Fukuoka-Zeki Sakura Park is located along the Kokai River near Fukuoka—Zeki (Fukuoka dam), one of 3 major historic dams in the Kanto region. This project was a collaboration effort between Ibaraki prefecture and Tsukubamirai city, to commemorate the merger of the town of Ina and the village of Yawara, which together form Tsukubamirai city. The park maintains and fully utilizes the vernacular landscape around the historic Fukuoka dam while paying careful attention to the ecological function of the site. Existing woods were preserved to the fullest extent possibly in order to maintain the biodiversity around the site.

Sakura(Cherry) theme was celebrated in the park concept, because the site was originally famous for its beautiful cherry promenade. The "Cherry and Wind" monument welcomes people at the main entrance, symbolizing the celebration of the birth of Tsukubamirai city. Providing an amenity where people can socialize and connect with nature and water was one of the important goals for this project. "Water Monument", "Mist Fountain" and "Jabu-Jabu Pond" are central park elements providing visitors with highly interactive water activity during summer. Seasonal changes such as cherry blossoms in spring, water amusement in summer, and fall color in autumn add year-round interest to the park.

福冈堰樱花公园坐落于福冈堰（福冈坝）附近的Kokai河畔，福冈堰是Kanto地区最具历史意义的三大堤之一。这是茨城县和筑波未来市的合作项目，以纪念组成筑波未来市的伊奈县与谷原县的合并。

公园保持且充分利用了具有历史意义的福冈大坝周围的民居景观，同时高度重视此地区的生态功能。为了保持该地区的生物多样性，现存的森林被进行了最大程度上的保护。

公园举办以“樱花”为主题的庆祝活动，这是因为该地区有史以来就闻名于她美丽的樱花漫步。在中心入口处伫立的“樱花与微风”纪念碑，欢迎着人们的到来，同时象征着欢庆筑波未来市的诞生。

这个项目的最重要的目标之一就是要给人们进行社交及亲近大自然和水提供便利。而水之源纪念碑、雾泉和Jabu—Jabu人工水池都是公园的中心设施，给游客提供了可在夏季有高度互动性的水上活动。

四季的变换，如春天的樱花绽放，夏日的水上嬉戏，秋日的落叶景观让公园一整年都充满了乐趣。

Shinagawa Chuo Park

品川中央公园

LOCATION: Tokyo, Japan
项目地点：日本 东京

AREA: 20,480 m^2
面积：20 480 平方米

COMPLETION DATE: 2007
完成时间：2007 年

LEAD DESIGNER: Tooru Miyakoda, FASLA
首席设计师：Tooru Miyakoda, FASLA

DESIGN COMPANY: Keikan Sekkei Tokyo Co., Ltd.
设计公司：Keikan Sekkei Tokyo Co., Ltd.

Shinagawa Chuo Park

Shinagawa Chuo Park, located in front of Shinagawa City Hall, features flexible plaza spaces capable of accommodating users for various events.

The Entrance Plaza, Central Plaza, and Multi—Purpose Plaza are designed to hold festivals of varying type and size. Symbolic fountains called "Mist Fountain" at the Entrance Plaza, and "Mountain Fountain" at the Central Plaza are connected by a canal, where children play freely in the water throughout the year.

One of the main design philosophies is "the beloved park is always visited". One of the primary functions of the park is disaster prevention. In the case of emergency the park, in conjunction with the ground floor of Shinagawa City Hall, is utilized as an evacuation and meeting point, as well as serving as a staging point for initial assessment and treatment.

To facilitate this function the design established a clear primary axis connecting the Entrance Plaza, Central Plaza, and Multi—Purpose Plaza, with the city hall building, which makes it easy for emergency workers and evacuees to grasp the space and route.

Important factors in the design include preservation of exiting camphor and cherry trees, as was providing comfortable spaces for visitors with fountains, canals, rock gardens, and lawn mounds. The rock gardens are wheel chair accessible. Play lots are equipped with excise play structures. All other elements such as planting, lighting and signs are designed to entertain the visitors. This park attracts a diversity of users such as families, couples, seniors, and children even on weekdays.

品川中央公园坐落于品川市政厅前，这里广场空间开阔，可以给使用者进行各种活动提供所需要的空间。

入口广场、中央广场和多用途广场分别设计，用于举办不同类型和规模的节日活动。入口广场的"雾泉"以及中央广场的"山泉"具有标志性，由一条水渠相连，那里一年四季孩子们都可以在水里嬉戏。

公园主要的设计理念之一就是"我爱的公园我常来"。公园主要功能之一就是预防灾害。万一有意外发生，与品川市政厅一楼相连的品川公园就可以用来当做疏散或集合点，以及充当一个初步评估和处理的周转站。

为了便于这一功能的实施，设计采用了一条清晰的主轴线，把入口广场、中央广场和多用途广场同市政厅大楼连在一起，使处理紧急事件的工作者和被疏散的人很容易弄清路线和方位。

设计的重要因素包括对现有的樟树和樱花树的保护，它们同喷泉、水渠、岩石园和草坪一起为游客提供了舒适的空间。岩石园轮椅也可以通行。游戏场地装配了用来锻炼的娱乐设施。其他的元素，如植物、照明和标志等都有设计来服务于游客。这个公园即使在工作日都吸引各式各样的游客，有家庭，夫妻，老人以及孩子。

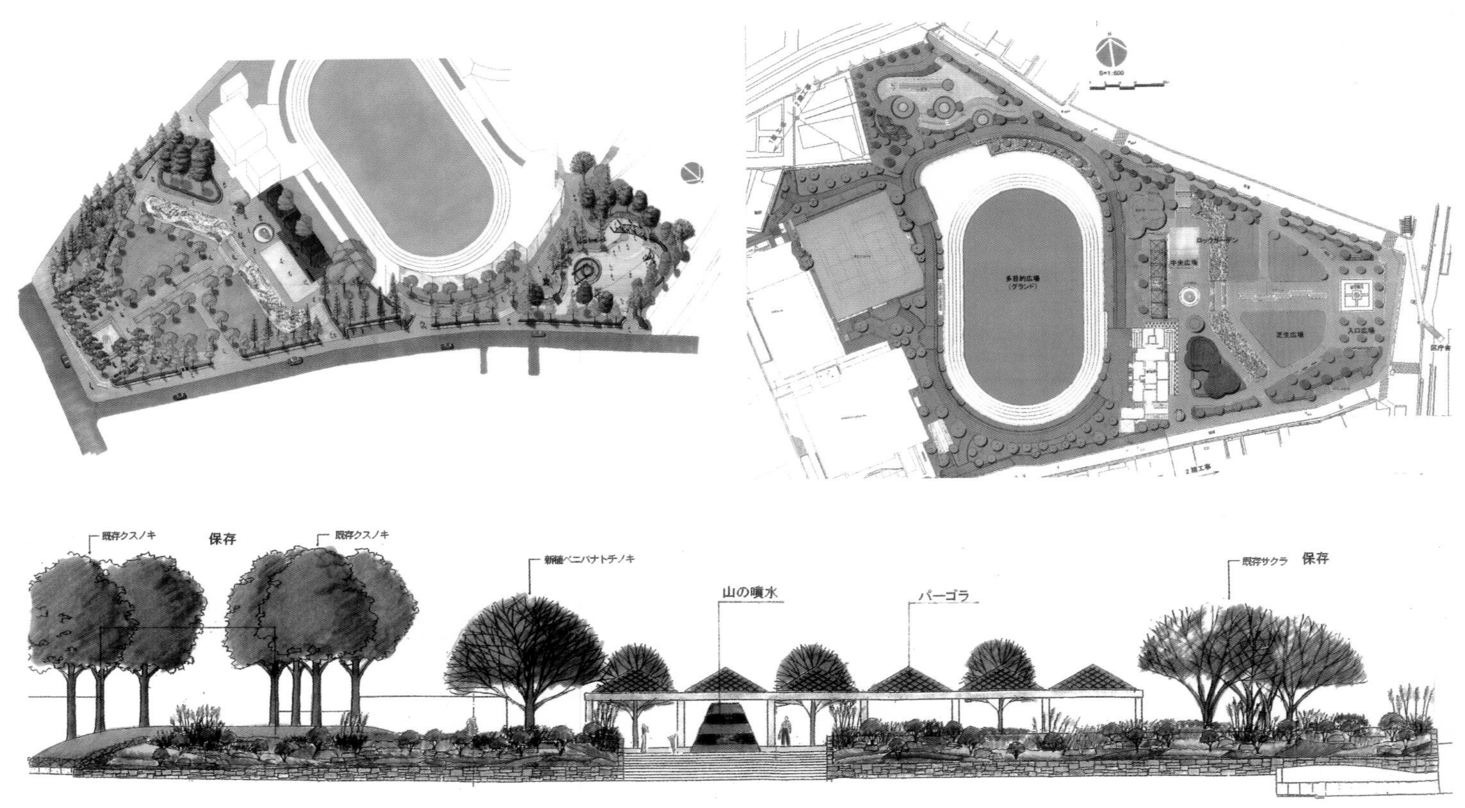
S=1:600
多目的広場
(グランド)
ロックガーデン
中央広場
芝生広場
入口広場
区庁舎
既存クスノキ
保存
既存クスノキ
新植ベニバナトチノキ
山の噴水
パーゴラ
既存サクラ
保存

品川区立
しながわ中央公園

CTOPOS DESIGN

EXPOSURE
The design from God's wisdom.
The design that exposes order of all things to make us believe in God.
Our design exposes the character and workmanship of God.
LIGHT
Light and spirituality stay together.
Light contains sun,moon, and stars.
Our design celebrates sun, moon, and stars' banquet of light.
Light adds value and life to materals, textures, and colors.
Light evokes emotion. Light makes all darkness disappear.
WATER
Water is the source of life.
God provides water through rain, fog, and clouds.
Water transforms its shape according to the space, and possesses the purity of always flowing from top to bottom.
Our design uses water to revive surrounding things.
LAND
Our design vitalizes everything on land.
Many are underneath land.
We design to serve and support land.
COLOR
God gave us color as a gift.
He gave color to make only humans feel.
Color is life itself.
That is why we design in the love of color.
INTIMACY AND SCALE
Purpose of our lives is to create relationships.
We enter into a relationship through a medium.
Relationship has to be intimate.
Intimacy is deeply connected to distance.
Our design is about detailed and friendly design of intimacy with a great sense of scale.
COMMUNICATION
Good communicaton is always great.
People and scape
Material and material
Wind,light,rain,earth and scape
Our design makes these things communicate with each other.
JOY
God told us to be joyful always.
Our design makes people feel full of joy.
LIFE
We want to depict peoples' lives in space.
Our design is not about our own thoughts or design.
Our design is about painting your lives in your space.

West Seoul Lake Park

西首尔湖公园

LOCATION: Seoul, Korea
项目地点：韩国 首尔

LEAD DESIGNER: Sehee Park; Chihun Kim, Hyejin Cho, Youngsun Jung, Saeyoung Whang, Miran Lyu
首席设计师：Sehee Park; Chihun Kim, Hyejin Cho, Youngsun Jung, Saeyoung Whang, Miran Lyu

LANDSCAPE ARCHITECT: Prof. Byunglim Lyu
景观设计师：Prof. Byunglim Lyu

TEAM: Junsuk Bae, Changwon Lee, Seong ki Kim, Taeyoung Ko, Sangkook Lee, Dongwon Kim, Suhyun Kim, Jungun Choi, Sanghun Yoon, Kwangho Hong, Hyunjung Lee, Myungbo Son, Heejin Park, Jihwan Kim, Semin Oh, Eunji Kim, Yoonyoung Lee, Yoon Jang, Wonki Jang
设计团队：Junsuk Bae, Changwon Lee, Seong ki Kim, Taeyoung Ko, Sangkook Lee, Dongwon Kim, Suhyun Kim, Jungun Choi, Sanghun Yoon, Kwangho Hong, Hyunjung Lee, Myungbo Son, Heejin Park, Jihwan Kim, Semin Oh, Eunji Kim, Yoonyoung Lee, Yoon Jang, Wonki Jang

DESIGN COMPANY: Ctopos Design
设计公司：Ctopos Design

West Seoul Lake Park

Our core design concepts were regeneration, ecology, and communication. We intended to redesign the boundary, which had disconnected park and local community. Approach to the park would be easier and our intervention would foster connectivity with local residents.

我们的核心设计理念为再生、环保及沟通。旧边界阻隔了公园和社区，我们将重新设计。新公园将更方便进入，我们的设计会加强其与当地居民的联系。

First, the park was created as an "open cultural art space", embodying the diversity of the area's identities and urban cultures, using the native environment to foster a self-organizing cultural zone for everyone.

首先，公园被打造成"开放的文艺空间"，既体现出地区特色和城市文化的多样性，又利用当地环境为公众打造以自我组织为特点的文化圈。

Second, the park preserved the existing natural topography and scenery to create a space for "urban ecology", integrating nature, culture, and urbanity. We engaged environmental elements of the site to make a cultural space for events with a home-grown culture—a scene open to all.

其次，为了结合自然、文化和都市风格，打造"城市生态"空间，园区保留了现有的自然地形和景观。我们利用该地区的环境因素，为本土文化活动打造空间，并向公众开放。

Third, it was created as a "people's park", uniting visitors by featuring a variety of abundant park events and special programs. This park is a citizen's park, citizen-generated by participation and communication. The school of ecology education will teach people the value of water and forest of the natural environment, scenery conservation, and nature study. Our use of program and context encourages all classes to communicate with one another.

第三，公园将以各种丰富的活动和特色节目聚拢游客，使之成为名副其实的"人民公园"。它是属于老百姓的，在其间每个人都可以参与和交流。该公园还是开展生态教育的学校，它告诉人们自然环境中的水和森林，风景保护和自然研究的价值。我们的项目和环境鼓励不同社会阶层的沟通。

Fourth, as a former water treatment plant reborn as a "site for urban culture", materials from the old plant were reused in surprising, inventive ways to transform raw nature into a new eco-functional space.

第四，公园的前身是一家净水厂，现在脱胎换骨成为"城市文化基地"，旧厂房的材料被重复利用，用令人称奇的、创造性的方法将原始的自然改造成为新的生态功能空间。

SITE PLAN of West Seoul Lake Park
Purification Pond
Long DinningTable for 100 People
Entry Plaza
Recycled Garden
Open Field
Playground
Terraced Garden
Sound Fountain responding to Aircraft Noise
Sub Gate
Sculpture Garden reuse of the existing structure
Art Deck
Visitor Center
N
0m 25m 50m 100m 200m

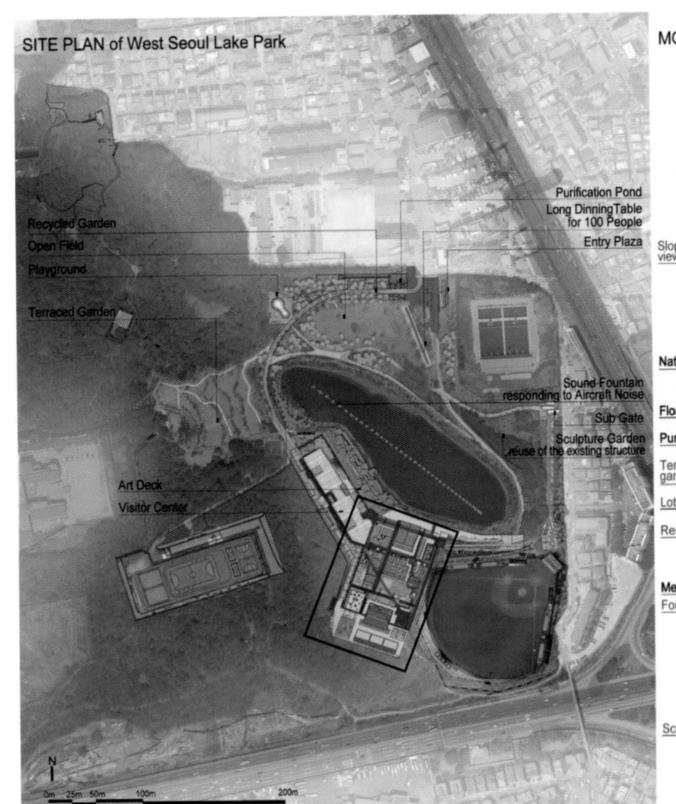

MONDRIAN PLAZA

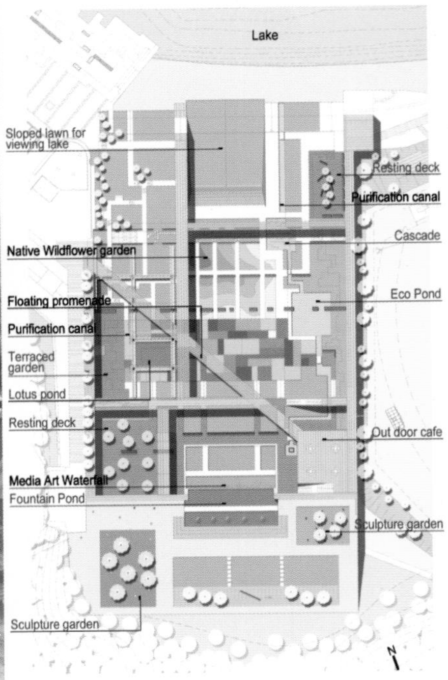

Lake
Sloped lawn for viewing lake
Resting deck
Purification canal
Cascade
Native Wildflower garden
Eco Pond
Floating promenade
Purification canal
Terraced garden
Lotus pond
Resting deck
Out door cafe
Media Art Waterfall
Fountain Pond
Sculpture garden
Sculpture garden
N

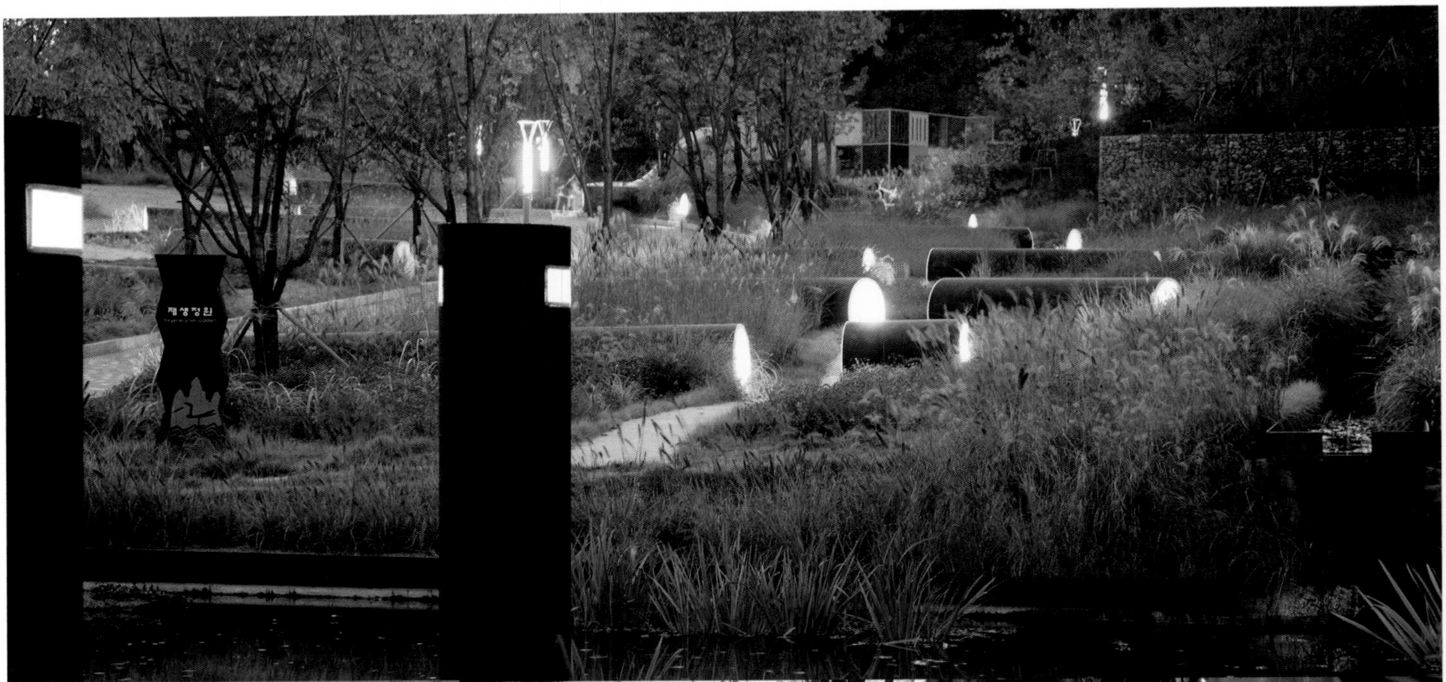

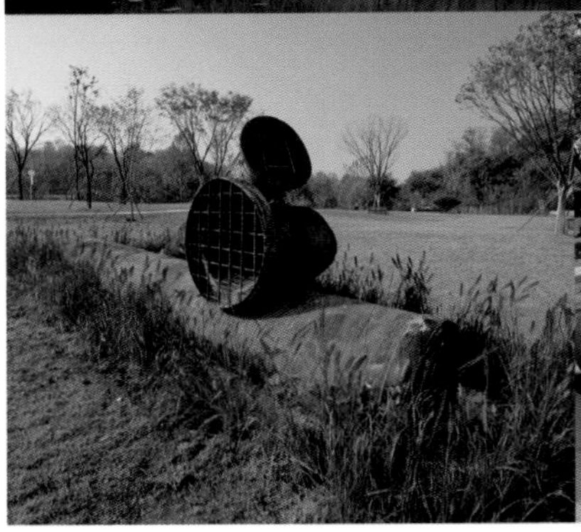

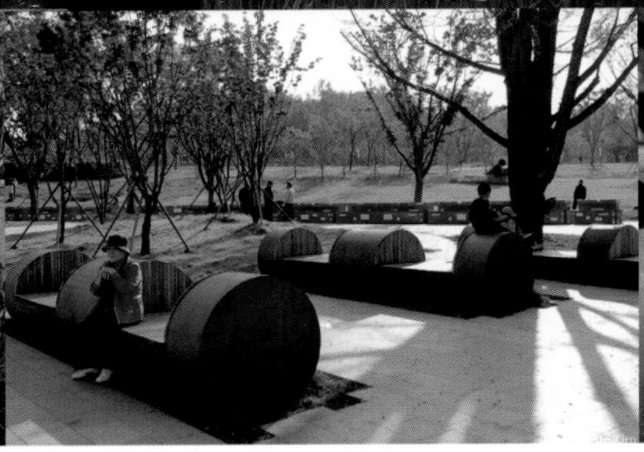

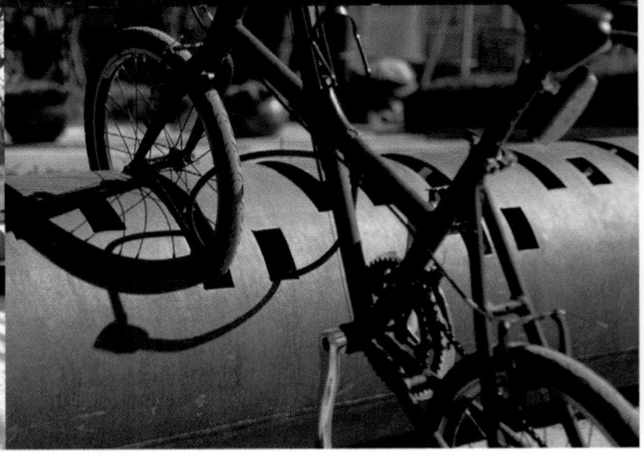

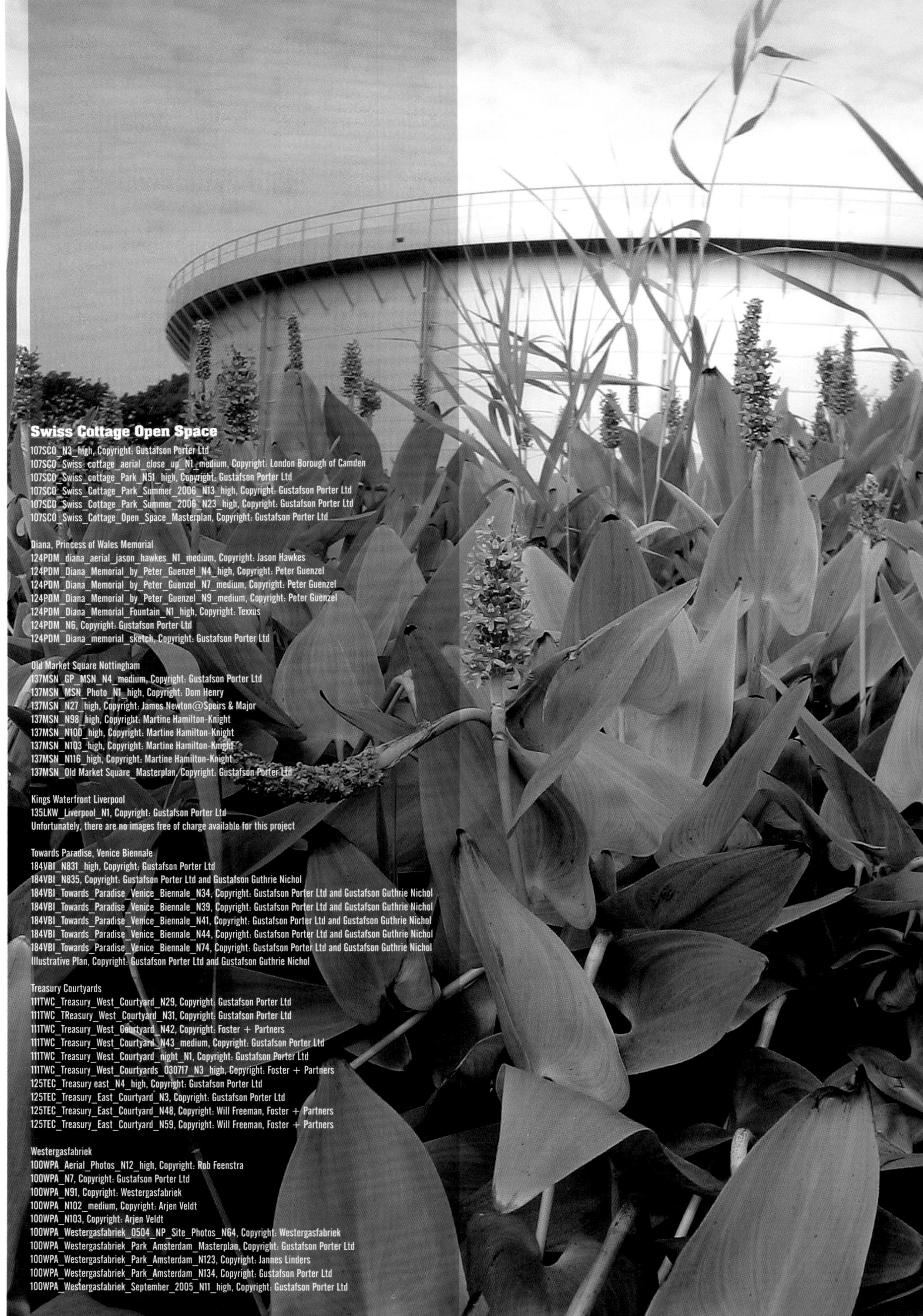

Swiss Cottage Open Space

107SCO_N3_high, Copyright: Gustafson Porter Ltd
107SCO_Swiss_cottage_aerial_close_up_N1_medium, Copyright: London Borough of Camden
107SCO_Swiss_cottage_Park_N51_high, Copyright: Gustafson Porter Ltd
107SCO_Swiss_Cottage_Park_Summer_2006_N13_high, Copyright: Gustafson Porter Ltd
107SCO_Swiss_Cottage_Park_Summer_2006_N23_high, Copyright: Gustafson Porter Ltd
107SCO_Swiss_Cottage_Open_Space_Masterplan, Copyright: Gustafson Porter Ltd

Diana, Princess of Wales Memorial
124PDM_diana_aerial_jason_hawkes_N1_medium, Copyright: Jason Hawkes
124PDM_Diana_Memorial_by_Peter_Guenzel_N4_high, Copyright: Peter Guenzel
124PDM_Diana_Memorial_by_Peter_Guenzel_N7_medium, Copyright: Peter Guenzel
124PDM_Diana_Memorial_by_Peter_Guenzel_N9_medium, Copyright: Peter Guenzel
124PDM_Diana_Memorial_Fountain_N1_high, Copyright: Texxus
124PDM_N6, Copyright: Gustafson Porter Ltd
124PDM_Diana_memorial_sketch, Copyright: Gustafson Porter Ltd

Old Market Square Nottingham
137MSN_GP_MSN_N4_medium, Copyright: Gustafson Porter Ltd
137MSN_MSN_Photo_N1_high, Copyright: Dom Henry
137MSN_N27_high, Copyright: James Newton@Speirs & Major
137MSN_N98_high, Copyright: Martine Hamilton-Knight
137MSN_N100_high, Copyright: Martine Hamilton-Knight
137MSN_N103_high, Copyright: Martine Hamilton-Knight
137MSN_N116_high, Copyright: Martine Hamilton-Knight
137MSN_Old Market Square_Masterplan, Copyright: Gustafson Porter Ltd

Kings Waterfront Liverpool
135LKW_Liverpool_N1, Copyright: Gustafson Porter Ltd
Unfortunately, there are no images free of charge available for this project

Towards Paradise, Venice Biennale
184VBI_N831_high, Copyright: Gustafson Porter Ltd
184VBI_N835, Copyright: Gustafson Porter Ltd and Gustafson Guthrie Nichol
184VBI_Towards_Paradise_Venice_Biennale_N34, Copyright: Gustafson Porter Ltd and Gustafson Guthrie Nichol
184VBI_Towards_Paradise_Venice_Biennale_N39, Copyright: Gustafson Porter Ltd and Gustafson Guthrie Nichol
184VBI_Towards_Paradise_Venice_Biennale_N41, Copyright: Gustafson Porter Ltd and Gustafson Guthrie Nichol
184VBI_Towards_Paradise_Venice_Biennale_N44, Copyright: Gustafson Porter Ltd and Gustafson Guthrie Nichol
184VBI_Towards_Paradise_Venice_Biennale_N74, Copyright: Gustafson Porter Ltd and Gustafson Guthrie Nichol
Illustrative Plan, Copyright: Gustafson Porter Ltd and Gustafson Guthrie Nichol

Treasury Courtyards
111TWC_Treasury_West_Courtyard_N29, Copyright: Gustafson Porter Ltd
111TWC_TReasury_West_Courtyard_N31, Copyright: Gustafson Porter Ltd
111TWC_Treasury_West_Courtyard_N42, Copyright: Foster + Partners
111TWC_Treasury_West_Courtyard_N43_medium, Copyright: Gustafson Porter Ltd
111TWC_Treasury_West_Courtyard_night_N1, Copyright: Gustafson Porter Ltd
111TWC_Treasury_West_Courtyards_030717_N3_high, Copyright: Foster + Partners
125TEC_Treasury east_N4_high, Copyright: Gustafson Porter Ltd
125TEC_Treasury_East_Courtyard_N3, Copyright: Gustafson Porter Ltd
125TEC_Treasury_East_Courtyard_N48, Copyright: Will Freeman, Foster + Partners
125TEC_Treasury_East_Courtyard_N59, Copyright: Will Freeman, Foster + Partners

Westergasfabriek
100WPA_Aerial_Photos_N12_high, Copyright: Rob Feenstra
100WPA_N7, Copyright: Gustafson Porter Ltd
100WPA_N91, Copyright: Westergasfabriek
100WPA_N102_medium, Copyright: Arjen Veldt
100WPA_N103, Copyright: Arjen Veldt
100WPA_Westergasfabriek_0504_NP_Site_Photos_N64, Copyright: Westergasfabriek
100WPA_Westergasfabriek_Park_Amsterdam_Masterplan, Copyright: Gustafson Porter Ltd
100WPA_Westergasfabriek_Park_Amsterdam_N123, Copyright: Jannes Linders
100WPA_Westergasfabriek_Park_Amsterdam_N134, Copyright: Gustafson Porter Ltd
100WPA_Westergasfabriek_September_2005_N11_high, Copyright: Gustafson Porter Ltd

GUSTAFSON GUTHRIE NICHOL LTD.

Gustafson Guthrie Nichol (GGN) is a landscape architecture practice based in Seattle, Washington. The 2011 recipient of the Smithsonian Cooper-Hewitt's National Design Award, GGN was founded in 1999 by partners Jennifer Guthrie, Shannon Nichol, and Kathryn Gustafson.

GGN's work is highly varied in scale and type—from furniture, like the Maggie Bench, to campuses and master plans, such as the Bill & Melinda Gates Foundation and the 2011 Downtown Cleveland Group Plan.

GGN offers special experience in designing high-use landscapes in complex, urban contexts. Boston's North End Parks, over the I-93 freeway, and Millennium Park's Lurie Garden, on a 5-story parking structure, are examples of GGN's designs for accessible, healthy, and sculptural urban spaces on rooftops and other urban structures.

The landform of each space is carefully shaped to feel serenely grounded in its context and comfortable at all times -- whether bustling with crowds, offering moments of contemplation, or doing both at once. The Robert and Arlene Kogod Courtyard, at the Smithsonian American Art Museum and National Portrait Gallery, is an example of a space designed to inspire and comfort either one or hundreds of people.

Project awards include multiple ASLA National Design Excellence Awards, Tucker Architectural Awards, and AIA/ASLA Honor and Merit awards for Design.

Cultuurpark Westergasfabriek

Westergasfabriek 文化公园

LOCATION：Amsterdam，The Netherlands
项目地点：荷兰 阿姆斯特丹

AREA：115,000 m²
面积：115 000 平方米

ARCHITECT：Francine Houben
建筑师：Francine Houben

PICTURES COPYRIGHT：Rob Feenstra，Gustafson Porter Ltd，Westergasfabriek，Arjen Veldt，Jannes Linders，
图片版权：Rob Feenstra，Gustafson Porter Ltd，Westergasfabriek，Arjen Veldt，Jannes Linders

DESIGN COMPANY：Gustafson Guthrie Nichol Ltd.
设计公司：Gustafson Guthrie Nichol Ltd.

Cultuurpark Westergasfabriek

The Westergasfabriek was a partially dismantled industrial site that retained vestiges of its original industrial and architectural layout. The buildings are a testament to the 19th century and the Industrial Revolution in which ideas of progress, evolution, and a better quality of life took a hold in Europe. It was called "the century of light" (enlightenment), in our times this ideology has evolved to create more sophisticated technologies and lighter industries.

Westergasfabriek 文化公园是一个部分拆除的工业遗址，保留原有的工业和建筑布局的痕迹。这些建筑见证了19世纪和工业革命时代，当时人们的想法是追求进步，追求革新和追求更高生活质量，这些想法成为欧洲当时的主导思想。它被称为“世纪之光”（启蒙运动），在我们这个时代这种思想已经有所演变，目的是为了创造更复杂的技术和轻工行业。

Amongst current cultural developments are an interest in leisure, entertainment, the environment and harmony in one's quality of life. Technology and urban pressures have eliminated the need and acceptability of a gas production facility in a dense urban setting. Gas production has thus been displaced and replaced by current cultural needs on sites such as the Westergasfabriek.

在当今的文化发展中，人们更加关注娱乐、休闲、生活环境的质量以及如何和谐相处。在人口稠密的城市中，技术和城市发展的压力使人们不再需要也不接受一个汽油生产设施，于是汽油的生产在像 Westergasfabriek 文化公园这样的场地上已经被迁移然后被当前文化的需要取代了。

Man's attitude towards landscape and the environment has also evolved. The site's location contains many features that are a testimony to this evolution. To the east is the historic Westerpark and the city centre, to the west the Overbracker polder, an area of re-established native vegetation. The new landscape design for the Westergasfabriek park seeks to illustrate in a contemporary form man's changing views and attitude towards the environment and its resulting landscape types. It also highlights the project's placement between city and nature.

人类对于风景和环境的态度也已经进步和发展了。项目所在地的许多特征也证明了这一点。花园的东边是具有历史意义的维斯塔公园和城市中心，西边是 Overbracker 洼地，经过重建，上面长着当地的植被。Westergasfabriek 公园新的景观设计力求以现代主义的形式突出体现人类对于环境和景观正在改变的视角和态度。该项目重点强调城市与自然的合理安排。

At the east end the ambience of the park reflects the more formal traditional garden type. The central area reflects the post–war (50s–70s) attitude towards landscape as a support for sports, leisure and recreation. The north–west Overbracker polder reflects the recent past (70s–90s) which is representative of a need for a pure nature/ecology approach. The west end reflects current thinking, that environmental harmony must be achieved with man as a participating partner.

在东侧的尽头，花园的环境氛围体现了更加正式的传统花园类型。中心地区反映出第二次世界大战后（20 世纪 50–70 年代）人们对景观的态度，人们更倾向于追求运动、休闲和娱乐。西北部的 Overbracker 洼地反映了刚刚过去的 20 年（20 世纪 70–90 年代），那个时代代表了人们对纯粹的自然和生态的需要。西侧的尽头反映出现代的思维方式：想要实现与环境和谐这一理念必须首先把人当成与环境和谐的参与者。

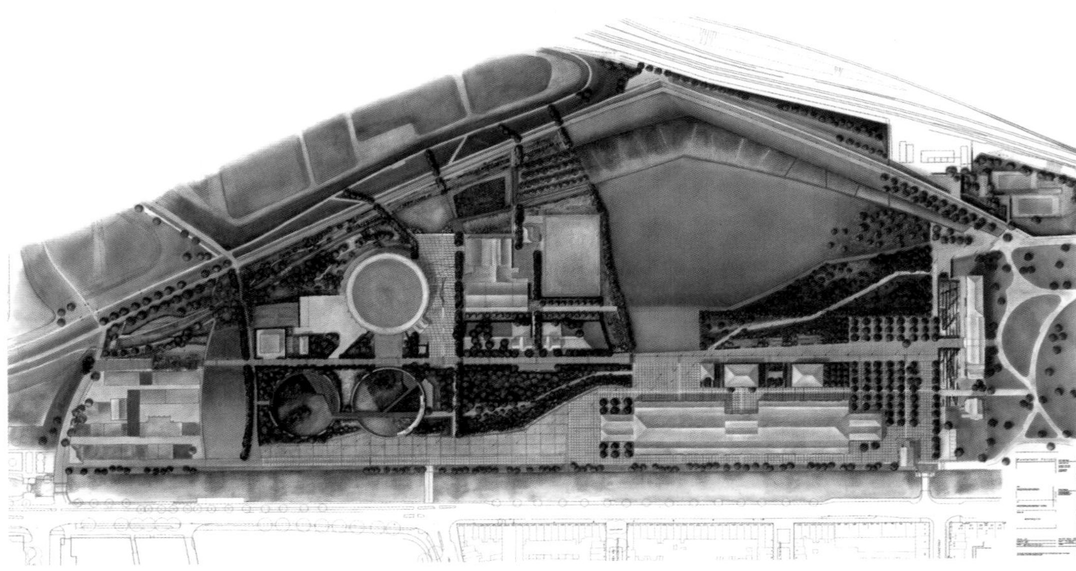

To achieve this balance on a heavily–polluted site required an understanding of the strategies required to allow the park's surfaces to be rendered safe for both people and plants. Polluted soil could not be taken off site to create new problems elsewhere, so a cut and fill balance was calculated bringing in new soil to displace polluted soil, retaining existing ground levels around the buildings and creating a new undulating terrain that was the consequence of surplus soil on the site.

为了在经历过严重污染的地区实现这种和谐，必须理解这些策略，必须允许公园外观看起来呈现出对于任何植物安全的特点。受污染的土壤无法运走，即便可以运走，又会对其他地方造成污染，文化公园

为了实现这一平衡在一个污染严重的网站需要一个理解的策略的表面要求允许公园要渲染的人来说都是安全、植物。污染土壤现场不能被创造新问题在其他地方，所以减少计算并填写平衡带来新的土壤污染土壤取代，保留现有建筑物周围地面水平而创造一个新的地形起伏所产生的效应是引起土壤剩余的地点。

SKI

Old Market Square Nottingham

诺丁汉旧市集广场

LOCATION：Nottingham，UK
项目地点：英国 诺丁汉

AREA：11,500 m^2
面积：11 500 平方米

COMPLETION DATE：2007
完成时间：2007 年

PICTURES COPYRIGHT：Gustafson Porter Ltd，Dom Henry，Martine Hamilton Knight/Builtvision，James Newton
图片版权：Gustafson Porter Ltd，Dom Henry，Martine Hamilton-Knight/Builtvision，James Newton

DESIGN COMPANY：Gustafson Guthrie Nichol Ltd.
设计公司：Gustafson Guthrie Nichol Ltd.

Old Market Square Nottingham

The Old Market Square Nottingham is the city's guardian space, a safe haven, a place to regain energy, wait and meet friends, be diverted momentarily from one's daily routines and to experience spectacular and well-organized civic and cultural events.

诺丁汉旧市集广场是该市的守护空间，一个安全的避风港，一个重获能量的地方，在这里可以等待和约见朋友，可以片刻地远离日常的忙碌，可以体验盛大有序的民众和文化活动。

The Square is one of Britain's oldest public squares, with an 800-year history as a market place, and at 11,500 m^2, the second largest after Trafalgar Square. The formal 1929 design by T.C.Howitt did not serve the requirements of a progressive European city, and following an international design competition in 2004, the Jury unanimously selected the new contemporary design.

该广场是英国最古老的公共广场之一，作为集市已有 800 年历史，占地面积为 11 500 平方米，是仅次于特拉法格广场的第二大广场。1929 年 T.C. 豪伊特的正式设计不能满足一个发展中的欧洲城市的需要，因此在随后的2004年国际设计比赛中，评委会一致选出了新的现代的设计。

The brief was to: provide unhindered access for all, use high-quality materials, provide new water features, introduce soft landscaping, integrate street furniture, create flexible performance space, allow people to linger, encourage 24-hour use, enable perimeter activity to spill out into the space, and attract pedestrians by virtue of its design. It also had to create a sense of place and reinforce the distinctive qualities and character of Nottingham.

大体构想是，提供畅通无阻的通道，使用高质量的材料，提供新的水文景观，种植草木，加入街道设施，创建灵活的活动空间，让人留连忘返，鼓励全天 24 小时使用，使得周边活动可以延伸到这里，并凭借其设计来吸引行人。另外，它还要创造一个地域感，增强诺丁汉与众不同的品质和特点。

Terraces of coloured granite blocks delineate level changes and hint at the geological strata below the Square's surface. Their tapering forms accommodate rows of benches, planters and water events, and set around a large flat and unobstructed surface used for markets and city events.

彩色花岗岩块铺的各个平台描绘出层面的变化，并暗示着广场表层下方的地质层。它们呈平面锥形，容纳了成排的长椅、花盆和水景，围绕在一个用于市场和都市活动平坦通畅的广场周围。

The terraced water feature comprises a reflecting pool, waterfall, rills, jets and a scrim that can be switched off and enables its use as an amphitheatre. All lighting is concealed, except for masts that create a range of lighting moods and support temporary trusses needed for events.

阶梯式水景包括一个倒影池、瀑布、溪流、喷口和一个可以关闭的纱幕，能将这一区域用作一个圆形剧场。除了创造一系列灯光氛围和支撑活动需要的临时支架所用的灯杆以外，所有照明都隐蔽起来。

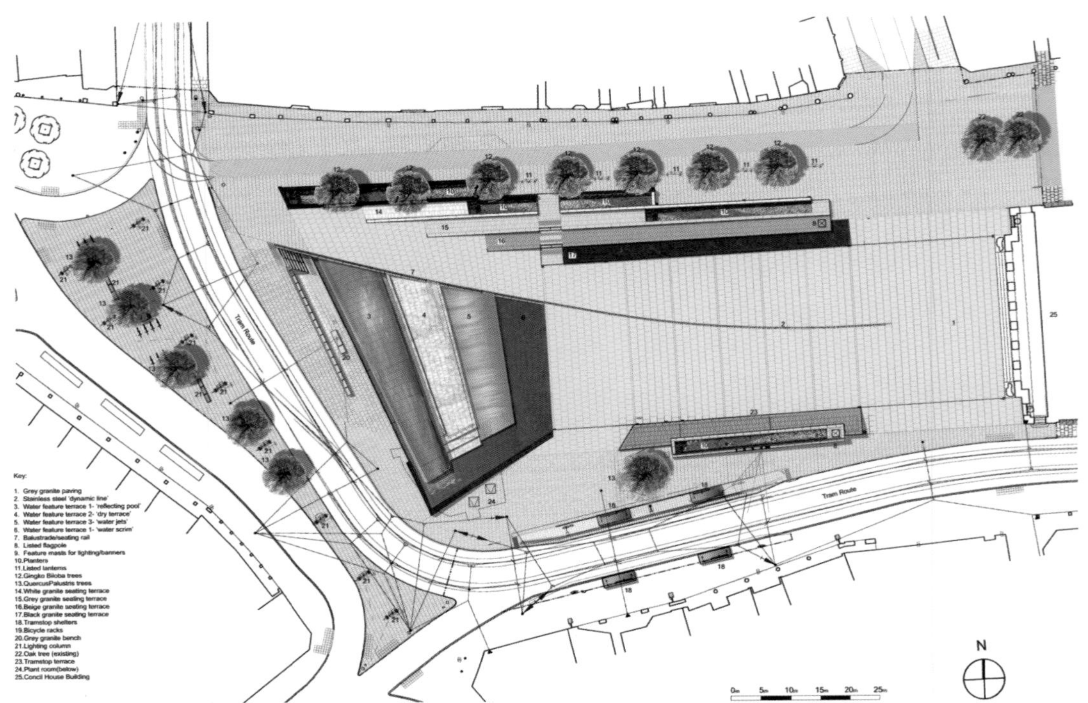

New direct, diagonal routes between Chapel Bar and Smithy Row, Long Row and Friar Lane leading to Nottingham Castle, enable access to the centre of the square and at the same time ensure pedestrians easy access to all parts of the city centre.

在教堂酒吧和铁匠铺街之间，长街和修士巷之间新的直接的对角线路，通往诺丁汉城堡，也通向广场的中心，同时确保行人方便到达城市中心的所有地区。

TECHNICAL PROFILE

The office uses the BIM program Archicad, and G–prog for building calculations. We use 3D Studio MAX, Rhino, and Maya for all visual presentations.

We use a variety of consultants in order to make our projects successful. In Bergen, we work closely with NODE AS for designing and drawing the concrete and steel in our buildings. Our most prominent project with NODE AS was the Aurland Lookout for the National Tourist Roads project for Statens Vegvesen. This project won the national construction prize. Recently we have used SWECO AS for concrete and steel work on the Solberg Tower and Park project for Statens Vegvesen in Sarpsborg.

Almost all of our building–approval paper work in Norway is done by NEAS AS. They have a staff of lawyers that complete all the necessary forms. Saunders Arkitektur AS has been reponsible for the design (Ark PRO) and control (KPR) for all of our projects in Norway. All of our buildings have received approval as we do a thorough job together with NEAS AS in determining the relevant building restrictions and limitations in the projects we design.

SAUNDERS ARKITEKTUR AS

Saunders Arkitektur AS is owned by Canadian Todd Saunders (born 1969 in Gander, Newfoundland) who has lived and worked in Bergen, Norway since 1996. The firm has recently been incorporated. Todd Saunders is the principal architect along with one employed architect, several apprentices, and an office manager. Todd Saunders has 15 years' experience (with S øk, PRO & KPR) while the other architects in the office have 2-8 years' experience.

Saunders has worked in countries such as Austria, Germany, Russia, and Latvia. Currently, and the office is working mostly in Norway, and also has projects in both Europe and Canada. The works of Saunders Arkitektur AS have been published in newspapers, magazines, and books worldwide. We have included a list of publication later in this presentation.

Todd Saunders received a Bachelor of Environmental Planning & Design from the Nova Scotia College of Art and Design, Halifax, Canada. While doing his bachelor, Saunders spent a semester as an exchange student at the Department of Architecture, Rhode Island School of Design. He subsequently received a Master of Architecture from McGill University, Montreal, Canada.

Saunders has been a part-time teacher at the Bergen Architecture School since 2001. He has lectured at various architecture and design schools in Scandinavia, Canada and England. In 2005 Saunders taught a one-week design/ build course in the International Architecture Program in Oulu, Finland. In the spring of 2006, Saunders was a visiting professor at the University of Québec in Montreal, Canada.

Todd Saunders is a member of NAL and is also a member of various architectural juries, including the Annual Norwegian Concrete Prize. He has been an architecture juror in Finland and Sweden. Todd Saunders has held various lectures for the Norwegian Architecture Association (NAL), Norsk Form, and at all 3 architecture schools in Norway.

Aurland Lookout

Aurland 瞭望塔

LOCATION：Norway
项目地点：挪威

SIZE OF THE RAMP：L x W x H ＝ 33.6m x 4.2m x 13.5m
坡道面积：长 x 宽 x 高＝33.6m x 4.2m x 13.5m

LANDSCAPE TEAM MEMBER：Todd Saunders & Tommie Wilhelmsen
景观团队成员：Todd Saunders & Tommie Wilhelmsen

ARCHITECT：Saunders Arkitektur & Wilhelmsen Arkitektur
建筑师：Saunders Arkitektur & Wilhelmsen Arkitektur

PHOTOGRAPHER：Todd Saunders
摄影师：Todd Saunders

DESIGN COMPANY：Saunders Arkitektur As
设计公司：Saunders Arkitektur As

Aurland Lookout

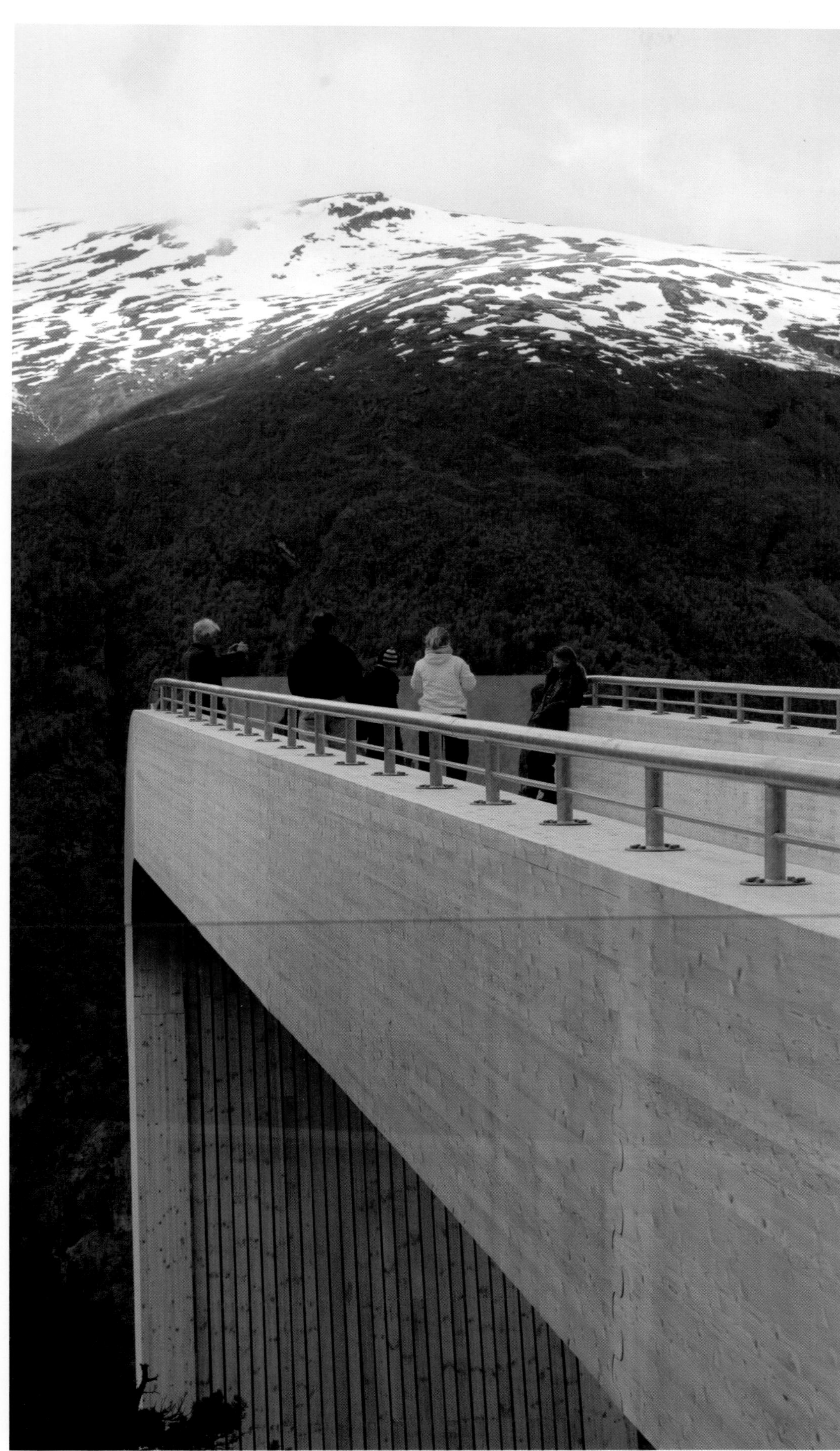

The ramp is constructed with load-bearing galvanized steel, and then covered with environmental pressure-reated pine. The load-bearing system is two parallell frames with a rectangular section. There are seven steel sections to make the on-site construction easier. The cross-section has been chosen to take into consideration and is 1,100 mm high, while the the "legs" of the ramp are a smaller at 300mm x 800mm. The most tension is in the curve of the ramp. A framework in the floor of the ramp criss-crosses from this point all the way to the foundation in the road. The floor of the ramp has an inner construcion of steel trusses with a centre distance of 1m.

坡道用承重的镀锌钢建造，上面铺有经过压力处理的环保松木。承重系统由带有一个矩形横截面的两个平行框架组成。现场有七根型钢，使施工更加方便可行。设计时已考虑到了横截面的问题，其高度为 1 100 mm，而坡道的"支腿"要小一些，尺寸是 300 mm × 800 mm。坡道弯曲处所受的张力最大，坡道地板内的构架在这里交叉，并一直延伸到道路的地基中。坡道的地面内部采用钢桁架建造，中心距为 1 m。

The floor of the ramp is 65 mm massive wood, placed with a small fall to each side of the ramp. The floor is screwed from the underside to the steel frame. The "rails" on the edge of the ramp are covered with 65mm of laminated wood. The wooden joints on the side of the rails are semi-cirle joints that make the joints less visible from a distance. The underside of the ramp is coved in open wooden slats.

坡道的地板用 65 mm 的厚重木材建造而成，在坡道的每侧都稍微向下倾斜。地板从下面用螺丝拧在钢架上。坡道边缘的"围栏"处铺有 65 mm 的胶合板，围栏一侧的木接头是半圆形的接头，从远处几乎看不到。坡道的内侧向木板条内凹陷。

The ramp has two concrete foundations. The upper foundation deals with most of the side forces from the wind. This foundation is made as a lying "U", and nine long tension bolts are fastened to the mountain under the road. The second, and lower foundation, is the foundation for the "legs" of the ramp. This foundation is bolted in two places to the mountain-side.

坡道有两个混凝土地基。上面的地基可以抵御大部分侧向风力，呈横“U”形，用九个长长的拉紧螺栓固定到公路下面的山上。下面的基础是坡道“支腿”的基础，在靠山的一侧有两处用螺栓固定。

The WC building is constructed of place-formed concrete that forms the floor, side-walls, and roof. The concrete is then covered with "elite-crete" to give a black coating of rubber-like surface. The end-walls are 65 mm solild wood. The end walls towards the fjords are mostly glass.

卫生间的地面、侧墙和屋顶都采用现场浇筑的混凝土建造，混凝土上面涂了一层黑色的“elite-crete”涂料，形成橡胶一样的表面。端墙采用 65 mm 的实木，朝向峡湾一侧的端墙则采用玻璃。

SIMONE AMANTIA SCUDERI

Born in Catania in 1974, he graduated on Agricultural Science and Technology and obtained a Master's Degree in Landscape Architecture with a thesis in the water garden. From 2005 he lives in Rome where he has been working as an agronomist and landscape architect. He has designed several public works in urban areas with the landscape architect Maria Cristina Tullio. He has participated in international competitions with Uta Zorzi Muhlmann.

Roof Gardens

屋顶花园

LOCATION：Rome，Italy
项目地点：意大利 罗马

AREA：3,200 m^2
面积：3 200 平方米

DESIGNER：Simone Amantia Scuderi
设计师：Simone Amantia Scuderi

Roof Gardens

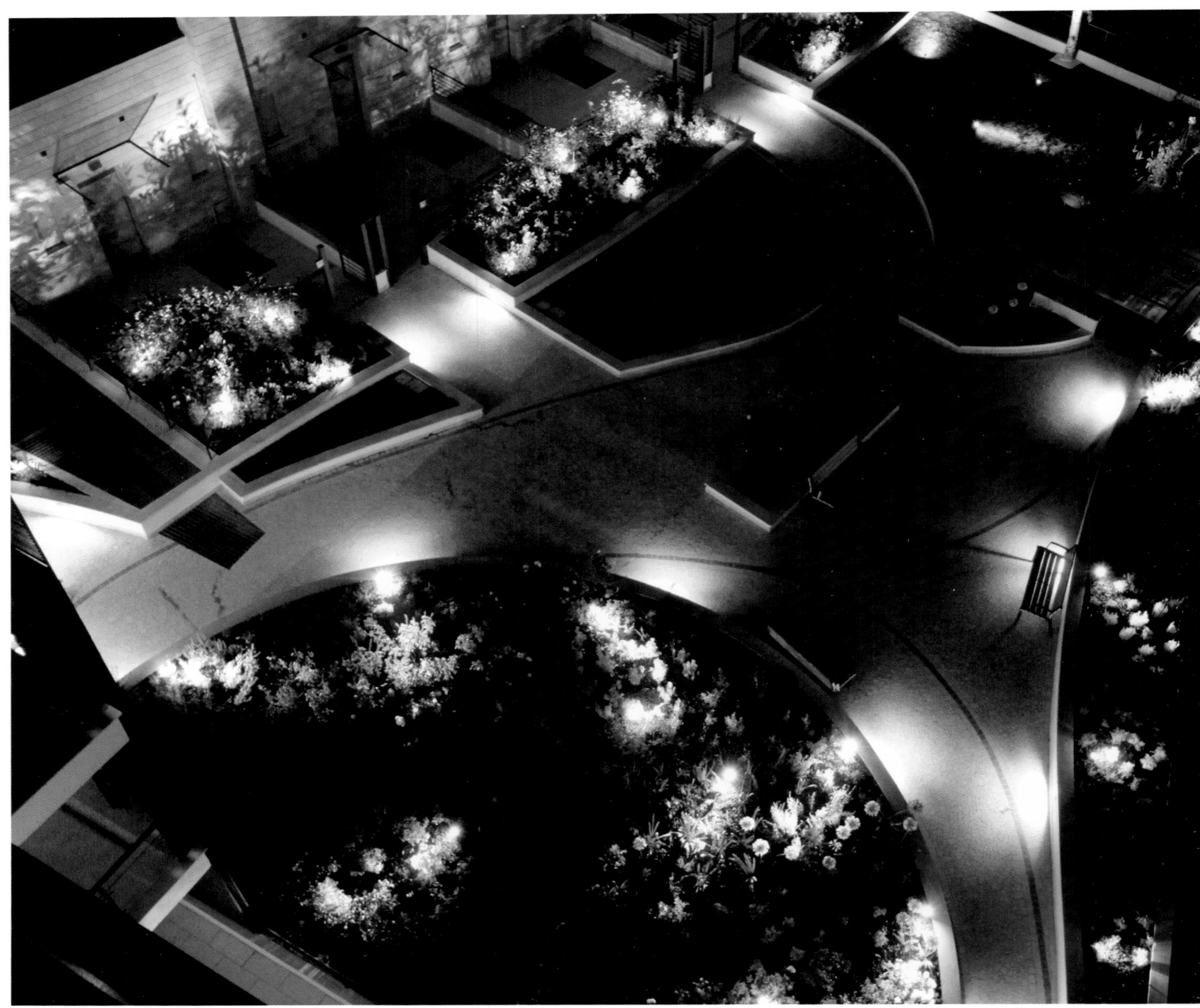

The site is located in a hilly area north of Rome, overlooking the wooded valley of the Rimessola. The building has two terraces, used as roof gardens, placed at different heights. The position of buildings on the upper terrace creates a courtyard enclosed with a splendid glimpse over the valley while the lower terrace is occupied by buildings only on two sides, opening the view on the Nature Reserve of the Insugherata. The two roof gardens have different characters and functions. one, with reception function, has a path leading to a loggia and features a green wall, a wall of water and a fountain. Passing the lodge we enter the first court that offers an intimate space closed with plant species that are able to create , chromatically and formally, a harmonious and unified space, with predominantly white blooms. The other terrace is characterized by the panoramic views and serves as a liaison between the houses and the pedestrian entrance to the basement garage. This space, facing west, is much more sunny and characterized by mediterranean species and bright colors. A dense hedge of Phyllirea angustifolia provides the necessary privacy for private spaces while the presence of flowering shrubs creates a pleasant garden to be crossed. A particular attention was paid to the construction techniques of the roof gardens and also to the visual impact of boundary high wall, solved by planting bamboo, hydrangeas and shade plants. Two green roofs, finally, characterize the coverage of the buildings; the garden, in fact, is meant also to be enjoyed from the upper floors, with different micro-landscapes: the roof of Sedum, the carpet of Iceberg roses, the water garden with leaf shape, the carpet of aromatic plants with gray leafs and blue blooms, the carpet of red roses and the fifth of bamboo.

项目位于罗马北部的一个丘陵区，俯瞰树木繁茂的Rimessola山谷。建筑有两个平台，被用做屋顶花园，高度不同。上面平台上建筑物的排布围出一个庭院，有着可瞥见山谷的极佳视角，下面平台只有两边是建筑物，可以视野开阔地看到Insugherata自然保护区。两个屋顶花园各具不同的特点和功能。上面花园有接待功能，有一条小路通向一个凉廊，景致包括一道植物墙，一道水墙和一个喷泉。过了门房，进入了第一个院子，那里提供了一个温馨的空间，由各种植物围起来，在色彩上和外形上营造出一个和谐统一的空间，有很多白色的花朵。下面花园的特点是有开阔的景色，其作用是房屋和通向地下车库的步行入口之间的一个连接。这一空间面朝西，阳光充足，地中海各类植物和鲜艳的色彩是其特色。一道浓密的菲利芮沙枣篱笆为私人空间提供了必要的隐密性，而开花的灌木形成了一个可以穿过的怡人花园。屋顶花园的建筑技巧投入了特别的关注，竹子，绣球花和喜阴植物的栽植解决了高高围墙带来的视觉影响。于是两个绿色的屋顶成为建筑物覆盖的特色，实际上，这个花园也是为了可以从上面的楼层欣赏，带有不同的微景观：景天植物屋顶、冰山玫瑰、叶子形水园、灰页蓝花芳香植物地面，红玫瑰地面和占地五分之一的竹子。

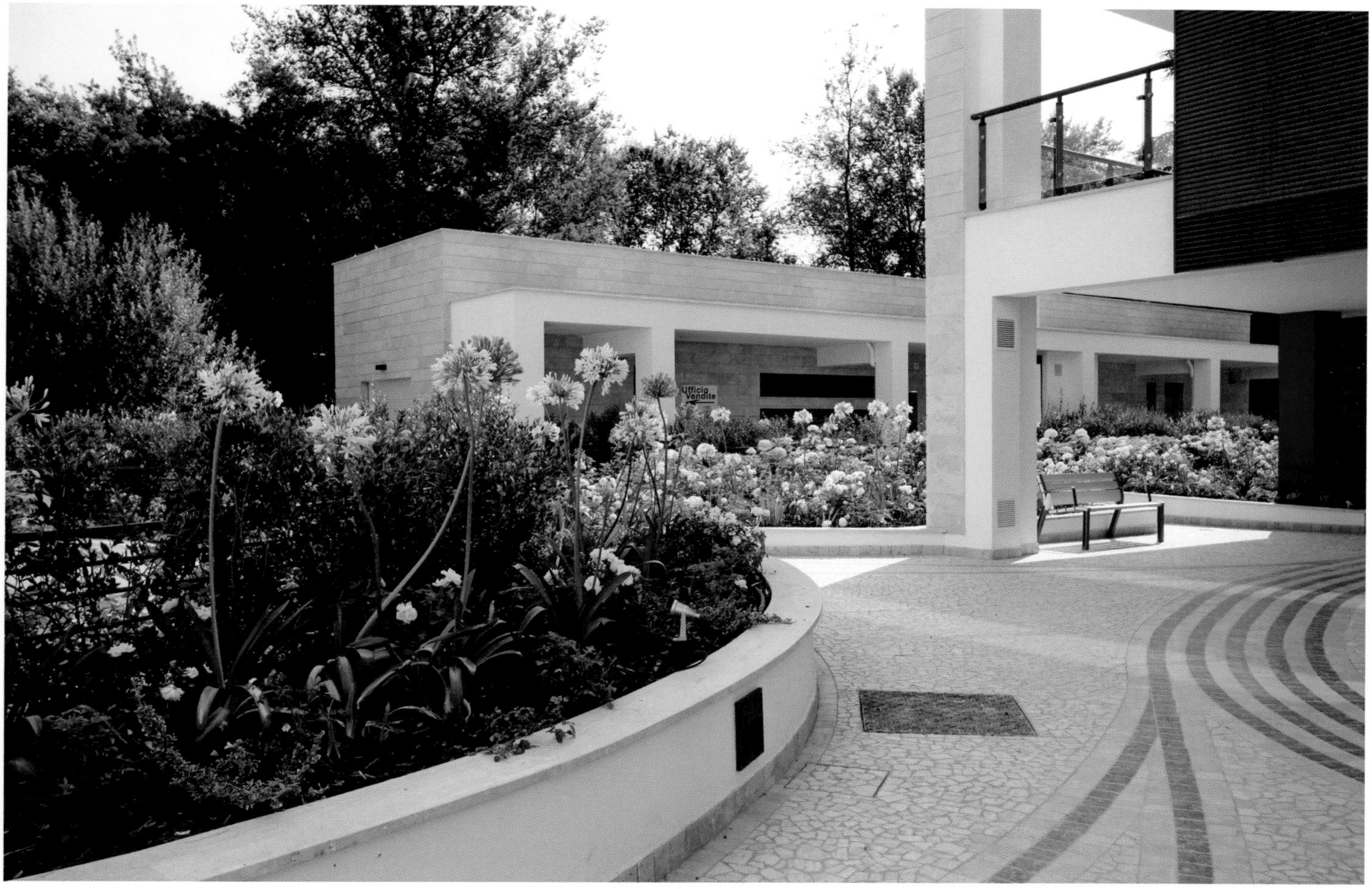
Ufficio
Vendite

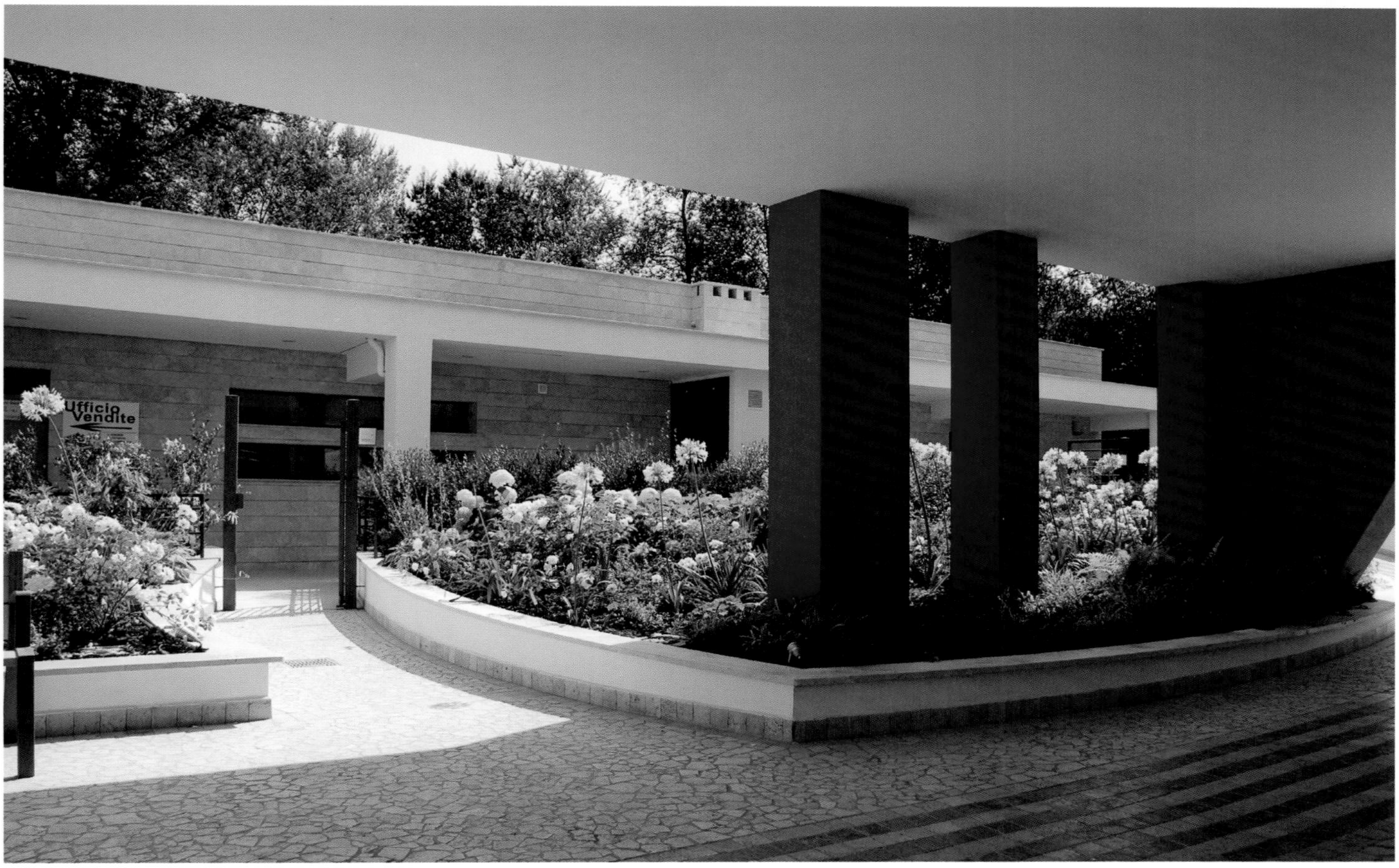
Ufficio
Vendite

ANTIQUES
148
02 Renewed Hope

BNIM

BNIM was founded 40 years ago with a commitment to design excellence and keen civic consciousness. Today, the firm contributes across a broad spectrum of topics with benefits felt locally and nationally.
We are designers. Thinkers. Collaborators. Architects. Planners. Landscape Architects. Friends. We are dreaming of the future. We are sustainable. We are BNIM.

BNIM and its employees have cultivated a culture that is more like a family than simply a place of employment. The firm has a horizontal hierarchy; everyone is encouraged to participate in the activities of the firm and push it forward. Weekly "Friday at 5" social gatherings and on-site Yoga classes bring people together in pursuit of health and fun. The annual design symposium, weekly all-staff meetings, regular breakfasts and pot-luck lunches cultivate a culture of friendship and a spirit of camaraderie that permeates the studio environment.

City of Greensburg Main Street Streetscape

风景如画的格林斯堡缅因街

LOCATION：Greensburg，USA
项目地点：美国 格林斯堡

COMPLETION DATE：2009
完成时间：2009 年

LEAD DESIGNER：Jim Schuessler，ASLA，Steve McDowell；Aaron Ross
首席设计师：Jim Schuessler，ASLA，Steve McDowell；Aaron Ross

DESIGN COMPANY：Bnim
设计公司：Bnim

City of Greensburg Main Street Streetscape

The City of Greensburg developed a downtown environment that not only provides a unique environment for residents and visitors, but also provides creative features that capture and recycle stormwater. This project is a part of an overall sustainable environment that was planned for the downtown business district. All components from planting and irrigation to seating, signage and materials are highly sustainable.

格林斯堡市不仅为居民和游客提供了独一无二的发达市区环境，还有可以循环再利用雨水的新设备。这是市中心商业区环境可持续发展总体计划项目的一部分。从植被和灌溉系统，到座椅及引导标志和这些设施及其材质，都是可以被充分持续利用的。

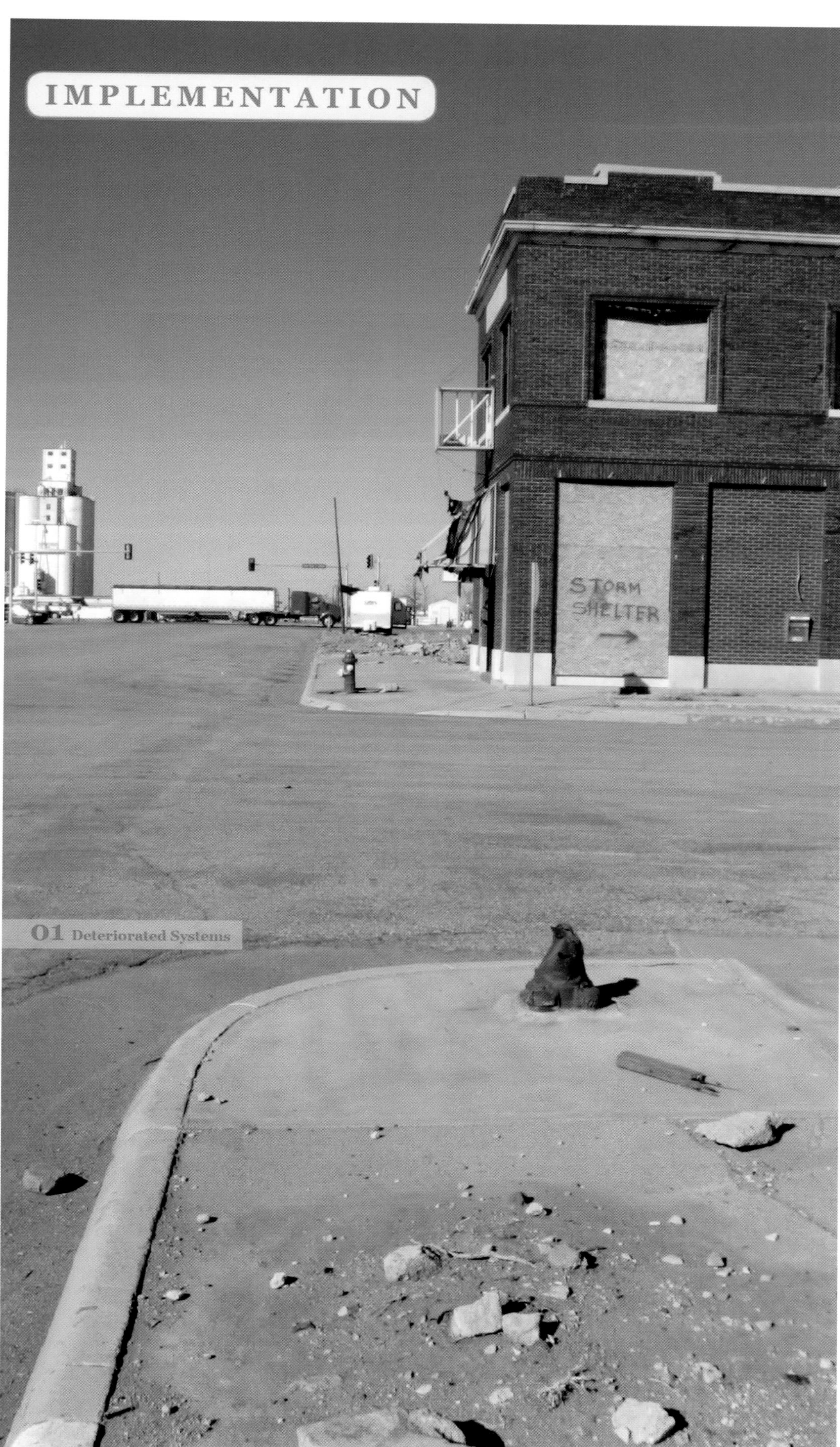

GREENSBURG DOWNTOWN STREETSCAPE

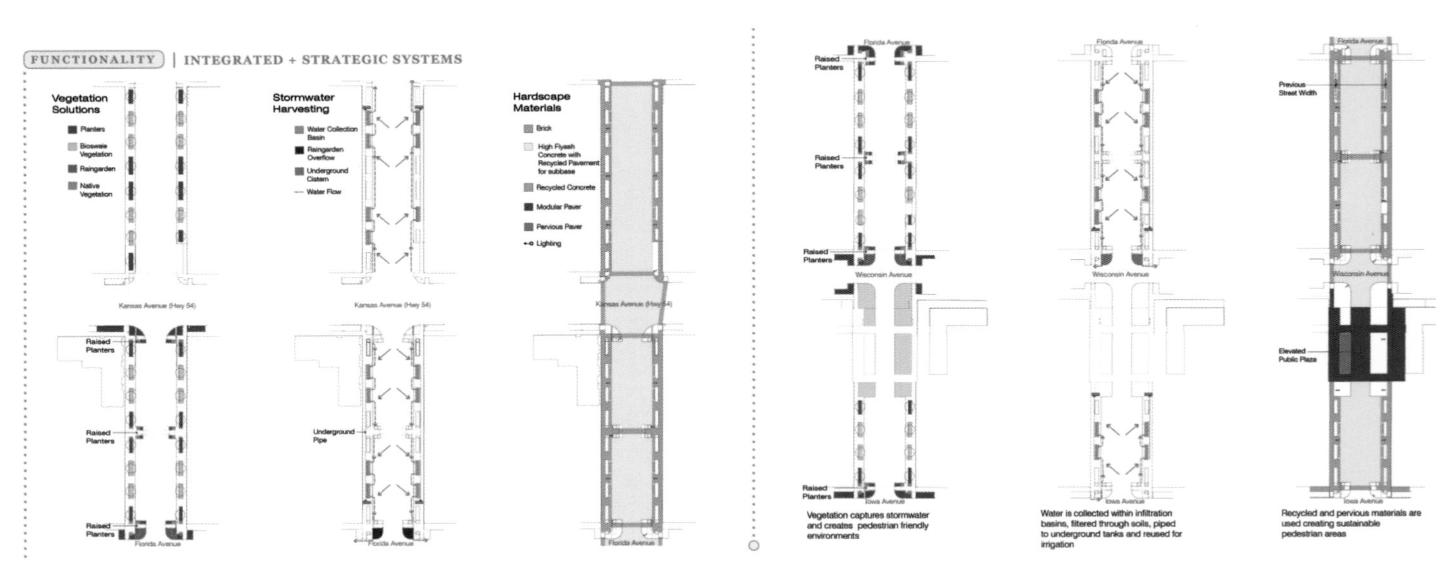

FUNCTIONALITY | INTEGRATED + STRATEGIC SYSTEMS
Vegetation Solutions
Planters
Bioswale Vegetation
Raingarden
Native Vegetation
Stormwater Harvesting
Water Collection Basin
Raingarden Overflow
Underground Cistern
Water Flow
Hardscape Materials
Brick
High Flyash Concrete with Recycled Pavement for subbase
Recycled Concrete
Modular Paver
Pervious Paver
Lighting
Kansas Avenue (Hwy 54)
Raised Planters
Underground Pipe
Florida Avenue
Wisconsin Avenue
Iowa Avenue
Previous Street Width
Elevated Public Plaza
Vegetation captures stormwater and creates pedestrian friendly environments
Water is collected within infiltration basins, filtered through soils, piped to underground tanks and reused for irrigation
Recycled and pervious materials are used creating sustainable pedestrian areas

$$$
Rain
Community
Environment
Family
Wind
growth
Prosperity

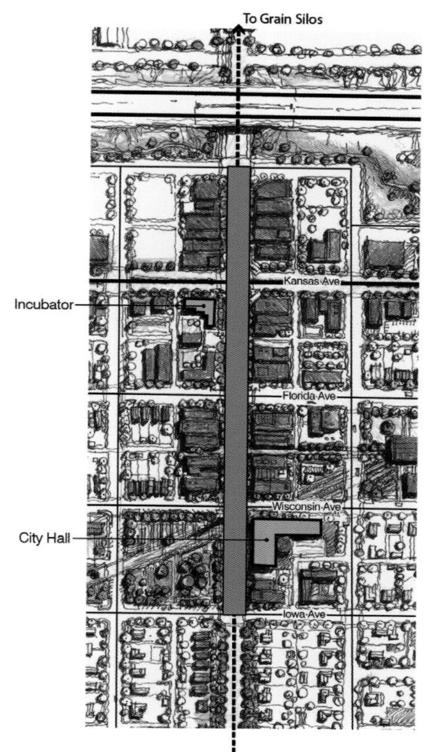

To Grain Silos
Kansas Ave
Incubator
Florida Ave
Wisconsin Ave
City Hall
Iowa Ave
K-12 School

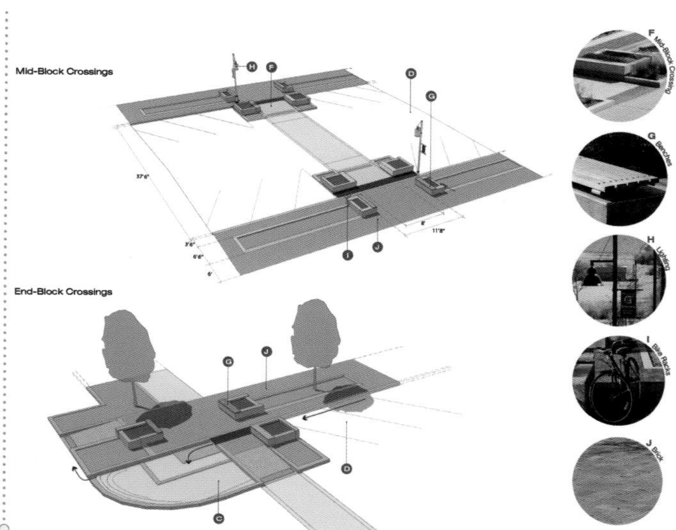

Mid-Block Crossings
End-Block Crossings

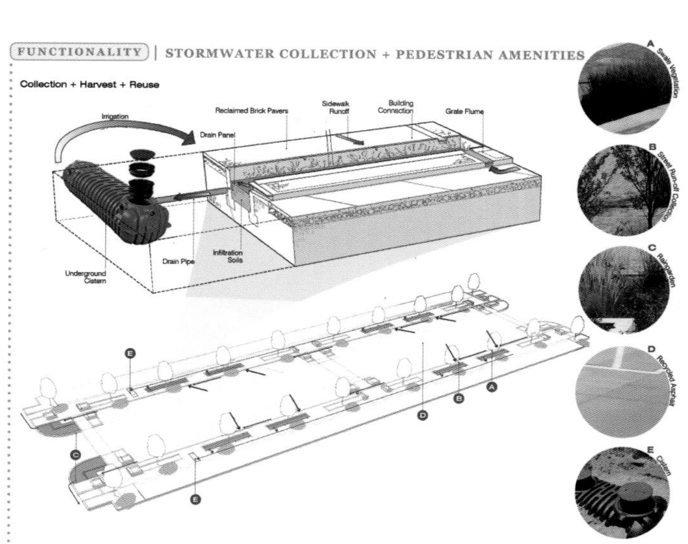

FUNCTIONALITY | STORMWATER COLLECTION + PEDESTRIAN AMENITIES
Collection + Harvest + Reuse
Irrigation
Drain Panel
Reclaimed Brick Pavers
Sidewalk Runoff
Building Connection
Grate Flume
Drain Pipe
Infiltration Soils
Underground Cistern

IMPLEMENTATION

06 Walkable Destinations

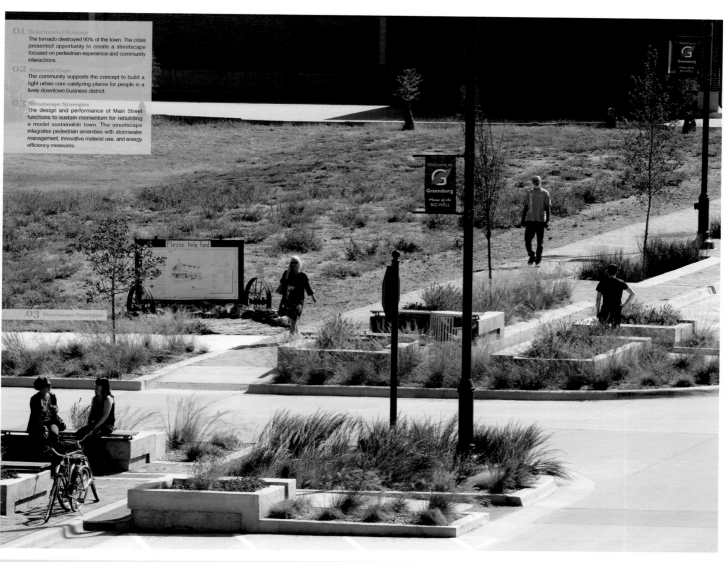

01 Deteriorated Systems
The tornado destroyed 90% of the town. The crisis presented opportunity to create a streetscape focused on pedestrian experience and community interactions.

02 Renewed Hope
The community supports the concept to build a light urban core catalyzing places for people in a lively downtown business district.

03 Streetscape Synergies
The design and performance of Main Street functions to sustain momentum for rebuilding a model sustainable town. The streetscape integrates pedestrian amenities with stormwater management, innovative material use, and energy efficiency measures.

03 Streetscape Synergies

08 Stormwater as Amenity
The raingardens and infiltration basins filter and collect stormwater which can be stored in underground cisterns. This greywater is used to irrigate streetscape plantings which serve as a significant pedestrian amenity.

09 Framework for Future Development
As one of the first major infrastructure projects in Greensburg, the Main Street Streetscape is a signature project that helps sustain momentum for rebuilding a model sustainable town. Many of the strategies integrated into the streetscape design serve as a model for additional city projects.

REED HILDEBRAND

CHRIS MOYLES

Chris Moyles is project manager and senior designer with eighteen years of experience. He is a graduate of the University of Virginia with a Masters' s in Landscape Architecture. Chris has taught at the Arnold Arboretum of Harvard University and the Boston Architectural Center. Prior to joining the firm, Chris was senior associate with Child Asociates Incorporated, where he managed projects for institutiongs including Massachusetts Institute of Technology, the New York Botanical Garend, Cornell University, the Mount—Edith Wharton Restoration Inc, andnumerous residential projucts. Chris is a membr of the American society of Landscape Architect in the state of Massachuseum; The parrish Aet Museum of Art; and long Dock at Beacon.

ADRIAN NIAL

Adrian Nial has 10 years experience as a Landscape Architect. He is a graduate of SUNY College of Environmental Science and Forestry an Syracuse University with a Bachelor degree in Landscape Architecture. Prior to joining Reed Hilderbrand, Adrian worked at Carolyn cooney and Associates and Paul Lu and Associats, where he managed and designed numerous recreational, institutional, and public projects. Current project involvement includes Syracuse University New Student Housing, Parrish Aet Museum, Silver Hill Hospital, and residential projects in Westchester County, NY.

REED HILDERBRAND

Saunders Arkitektur AS is owned by Canadian Todd Saunders (born 1969 in Gander, Newfoundland) who has lived and worked in Bergen, Norway since 1996. The firm has recently been incorporated. Todd Saunders is the principal architect along with one employed architect, several apprentices, and an office manager. Todd Saunders has 15 years' experience (with S øk, PRO & KPR) while the other architects in the office have 2-8 years' experience.

Saunders has worked in countries such as Austria, Germany, Russia, and Latvia. Currently, and the office is working mostly in Norway, and also has projects in both Europe and Canada. The works of Saunders Arkitektur AS have been published in newspapers, magazines, and books worldwide. We have included a list of publication later in this presentation.

Todd Saunders received a Bachelor of Environmental Planning & Design from the Nova Scotia College of Art and Design, Halifax, Canada. While doing his bachelor, Saunders spent a semester as an exchange student at the Department of Architecture, Rhode Island School of Design. He subsequently received a Master of Architecture from McGill University, Montreal, Canada.

Saunders has been a part-time teacher at the Bergen Architecture School since 2001. He has lectured at various architecture and design schools in Scandinavia, Canada and England. In 2005 Saunders taught a one-week design/ build course in the International Architecture Program in Oulu, Finland. In the spring of 2006, Saunders was a visiting professor at the University of Québec in Montreal, Canada.

Todd Saunders is a member of NAL and is also a member of various architectural juries, including the Annual Norwegian Concrete Prize. He has been an architecture juror in Finland and Sweden. Todd Saunders has held various lectures for the Norwegian Architecture Association (NAL), Norsk Form, and at all 3 architecture schools in Norway.

Berkshire Boardwalk

伯克希尔浮桥

LOCATION：Stockbridge
项目地点：斯托克布里奇

AREA：822.96 m
面积：822.96 米

LEAD DESIGNER：Reed Hilderbrand
首席设计师：Reed Hilderbrand

PRICIPAL IN CHARGE：Gary R. Hilderbrand，FASLA
主要负责人：Gary R. Hilderbrand，FASLA

Berkshire Boardwalk

The owners of this new 2,700-foot-long boardwalk describe the work as "a remarkable discovery" of a place that was previously unreachable and unknowable. Threading a narrow path through the edge of a 50-acre wetland adjacent to their home, a team of carpenters assembled the boardwalk completely by hand, in the water, without machines. The result has been deemed a successful habitat enhancement and an exemplary permitting precedent for the town. For the owners it is a lasting and unforgettable experience in all seasons.

这长达2 700英尺的新浮桥被业主形容为一块从前无人到达和了解的土地上“一个了不起的发现”。他们居所附近有个50英亩的湿地，一队木匠穿过这湿地边缘的一条狭长的道路，在水中完全靠手工组装木桥，没有借助任何机械。于是这里被认定是休养生息的好地方，并被批准做为模范镇的先例。对业主来说，无论在什么季节，这都将是一个永生难忘的地方。

The landscape architect was hired to assist the owners in their efforts toward greater stewardship of their property—primarily in the regrowth forest that covers 70% of the land. Though the owners were acutely aware of the larger benefits of land drainage, storage, and habitat associated with the wet potions of their landholding, they could not imagine the experience of being in the middle of the marsh—it was completely inaccessible.

业主们聘请景观设计师来更有效的管理自己的产业，主要是希望重新达到70%的绿化率。虽然业主能够意识到土地排水、存储及建立住所会有更大的好处，但还是不能接受生活在沼泽中是什么样子——不会有人愿意生活在其中的。

As part of ongoing efforts in woodlot management, edge restoration, and meadow extension, the landscape architect proposed to extend a circuit trail along the upland edge and then out over the water. The design team engaged a nine—month long review process with conservation biologists, permit specialists, contractors, the property manager, and conservation commissioners to ensure adequate protection of the resource and mitigation of limited construction disturbance. The case for Conservation Commission approval hinged on careful evaluation of hydrologic and biotic characteristics of the site, the use of low-impact technologies in construction, and specific design characteristics devised to enhance wetland habitat.

作为正在进行努力的林地管理一部分，景观设计师建议沿着高地边缘增加绿地来提高蓄水量。
设计团队包括生物环保学家、相关当局、承包商、物业经理和环境保护专员，他们事先做了一个为期9个月的审查，来减少施工所带来的破坏，以确保尽可能地保护自然资源。为了提高湿地栖息地保护委员会对其批准的可能性，对水文和生物特性进行了严格评估，采用低影响施工技术，并且根据具体的特点来制定设计方案。

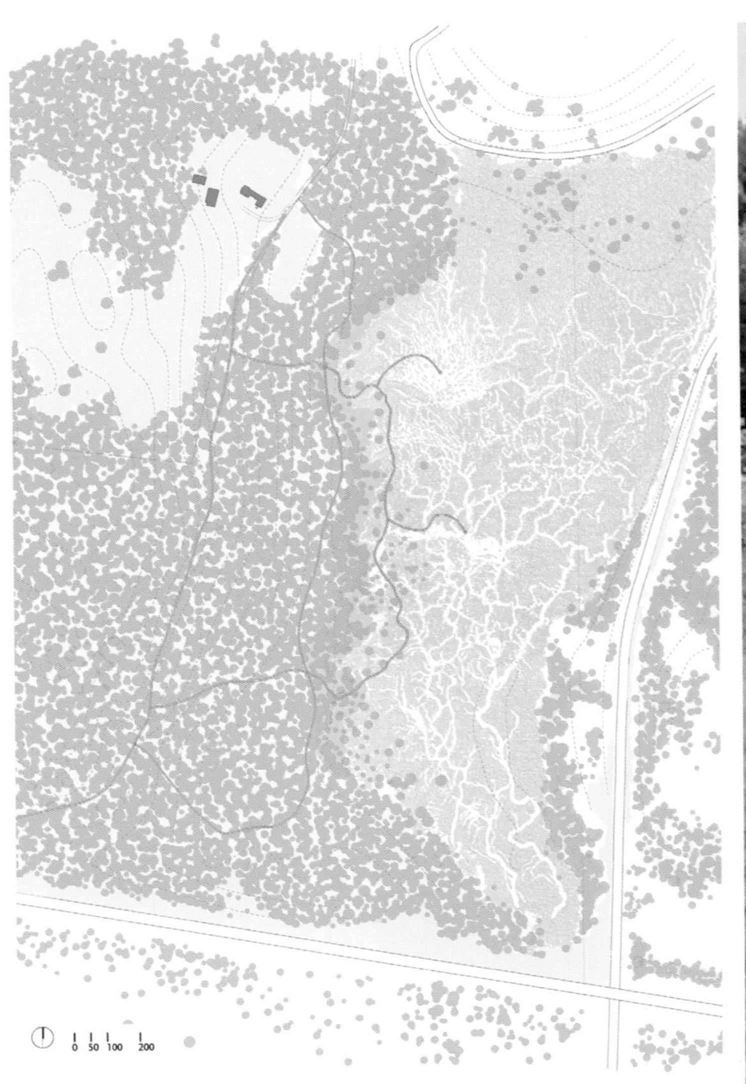
0 50 100 200

Warren T. Byrd Jr.
RLA, FASLA, Principal

As founding principal, Warren Byrd has led the firm in a wide range of public and private landscape projects throughout the United States, as well as Canada, New Zealand, the Netherlands, Brazil, Baja Mexico, Antigua, Russia, and China. Recent and current work includes the Flight 93 Memorial in Shanksville, PA with Paul Murdoch Architects of Los Angeles, CA; a two-block urban sculpture garden on the Gateway Mall in St. Louis, MO; the Dell stormwater park at the University of Virginia; the Asia Trail at the National Zoo in Washington, D.C.; the restoration of Montalto at Monticello in Albemarle County, VA; master landscape plans for Caltech in Pasadena CA, St. John's College in Annapolis MD, and the University of Miami FL in Coral Gables; the landscape master plan for Watercolor's 500-acre community on the Gulf Coast of Florida; the City of Lynchburg's urban waterfront park on the James River; and several residential projects in Virginia, North Carolina, and New York. The work of the firm has garnered over 50 national and regional design awards within the past ten years.

In addition to his 30+ years of practice, Warren taught for 26 years at the University of Virginia, serving as Chair of the Department of Landscape Architecture for seven years. His many honors have included an All-University teaching award, a CELA teaching award, and two Bradford Williams Medals for articles published in Landscape Architecture magazine. Now a Professor Emeritus, Warren's particular focus, in both his teaching and his practice, has been on the understanding and adaptation of natural systems and plant communities as they might best influence sustainable strategies of design. He received his B.S. in Horticulture from Virginia Tech in 1975 and his Master of Landscape Architecture from the University of Virginia in 1977. He is a Fellow of the American Society of Landscape Architects.

Thomas L. Woltz,
RLA, FASLA, Principal
Thomas Woltz began working with Warren Byrd in 1997 and became a partner of Nelson Byrd Woltz in 2004. During his time with the firm, Thomas has led designs of a broad range of institutional, and corporate projects in the US and abroad including The Peggy Guggenheim Sculpture Garden in Venice, Italy, Luckstone Corporation, Richmond VA, Washington and Lee University in Lexington, VA, The McIntire School of Commerce at the University of Virginia, Round Hill, Jamaica, and The National Arboretum of New Zealand, Eastwoodhill Arboretum. Thomas has also led design work on private gardens and farmland in a dozen states and New Zealand over 15 years of practice.

Thomas holds Masters degrees in Architecture and Landscape Architecture from the University of Virginia. Prior to graduate studies, he worked for five years in Venice, Italy where he developed an intense interest in architectural craft that continues to influence his design work. Thomas has held a part-time faculty position in the University of Virginia School of Architecture for 14 years teaching site planning and land analysis. More recently Thomas was instrumental in the establishment of the Conservation Agriculture Studio within Nelson Byrd Woltz. This is a family of projects that share information and seek to interweave sustainable agriculture with best management practices for conservation of wildlife, indigenous plants, soil and water. Currently the studio is dealing with more than 10,000 acres of cultivated land in VA, CA, NC, SC, KY, CT, NY, and New Zealand. Thomas serves on the Board of Directors of The Cultural Landscape Foundation and has been named to the 2011 ASLA Council of Fellows. He is frequently invited to speak around the country and internationally.

NELSON BYRD WOLTZ LANDSCAPE

NELSON BYRD WOLTZ LANDSCAPE ARCHITECTS
Founded in 1985, Nelson Byrd Woltz (NBW) is a 30-person landscape architecture firm with offices in Charlottesville Virginia and New York City. Committed to education and conservation, the firm has been involved in a broad array of public and private projects including public parks, botanic gardens and zoos, private gardens and estates, academic institutions, corporate campuses, and town planning. The firm actively seeks this diversity of project scales and types to cultivate the creativity of the professional staff. The firm's work has garnered over 70 national and regional awards and has been widely published.

Carnegie Hill House

卡内基山别墅

LOCATION：New York ，USA
项目地点：美国 纽约

LANDSCAPE ARCHITECT：Nelson Byrd Woltz
景观设计师：Nelson Byrd Woltz

TEAM：Nelson Byrd Woltz Landscape
设计团队：Nelson Byrd Woltz Landscape

Carnegie Hill House

Four gardens created in a constricted urban context provide a sanctuary to parents raising their children, pollinators and birds raising their young. This analogy to the "nest" provides an immersive learning experience in predominantly native plants connecting the owner to four seasons and an awareness of other species and their needs in an urban environment: water, habitat and forage. Details in planters, paving and furnishings draw inspiration from woven assemblage to reinforce this analogy of stewardship.

在拥挤的城市，四个花园简直就是父母们养育自己孩子，鸟儿虫儿养育自己后代的圣地。这个“巢”比喻了一个引人入胜的学习经验，即：水，土壤和肥料。业主四季都可以养殖当地主要植物，并且也可以根据他们的需求，在城市环境中养殖其他品种。栽种的细节，土壤的摊铺和布局，需要借鉴组合家具这一比喻来巩固管理。

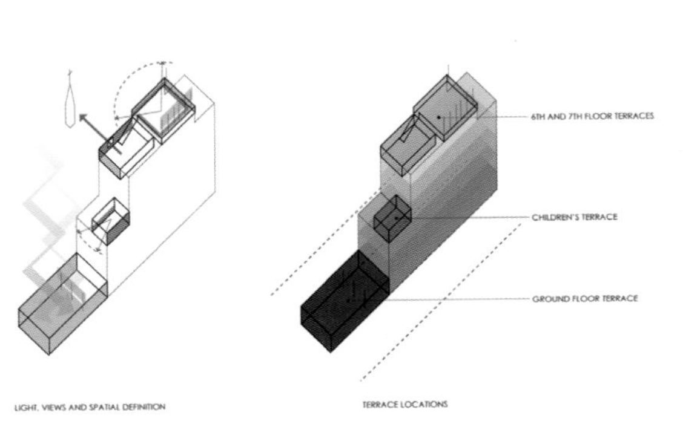

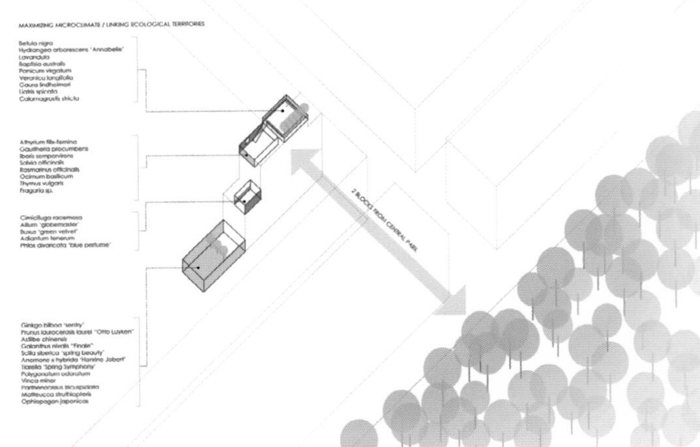

TEAK PLANTERS
BLUESTONE PAVING MIMICS CHURCH ROOF PATTERN
RIVER BIRCH
SLIDING TEAK SCREEN
SANDBOX
GREENWALL
SLATE AND TEAK SCREEN WALL
TEAK PLANTERS
SENTRY GINKGOS
LOCUST SLABS
REMNANT STONE PAVING FROM INTERIOR SPACE
BLUESTONE NEST PAVING
EXISTING BASIN AND FOUNTAIN

EAST 91ST STREET
7TH FLOOR TERRACE
6TH FLOOR TERRACE
CHILDREN'S TERRACE (FOURTH FLOOR)
GROUND FLOOR TERRACE
ADJACENT CHURCH

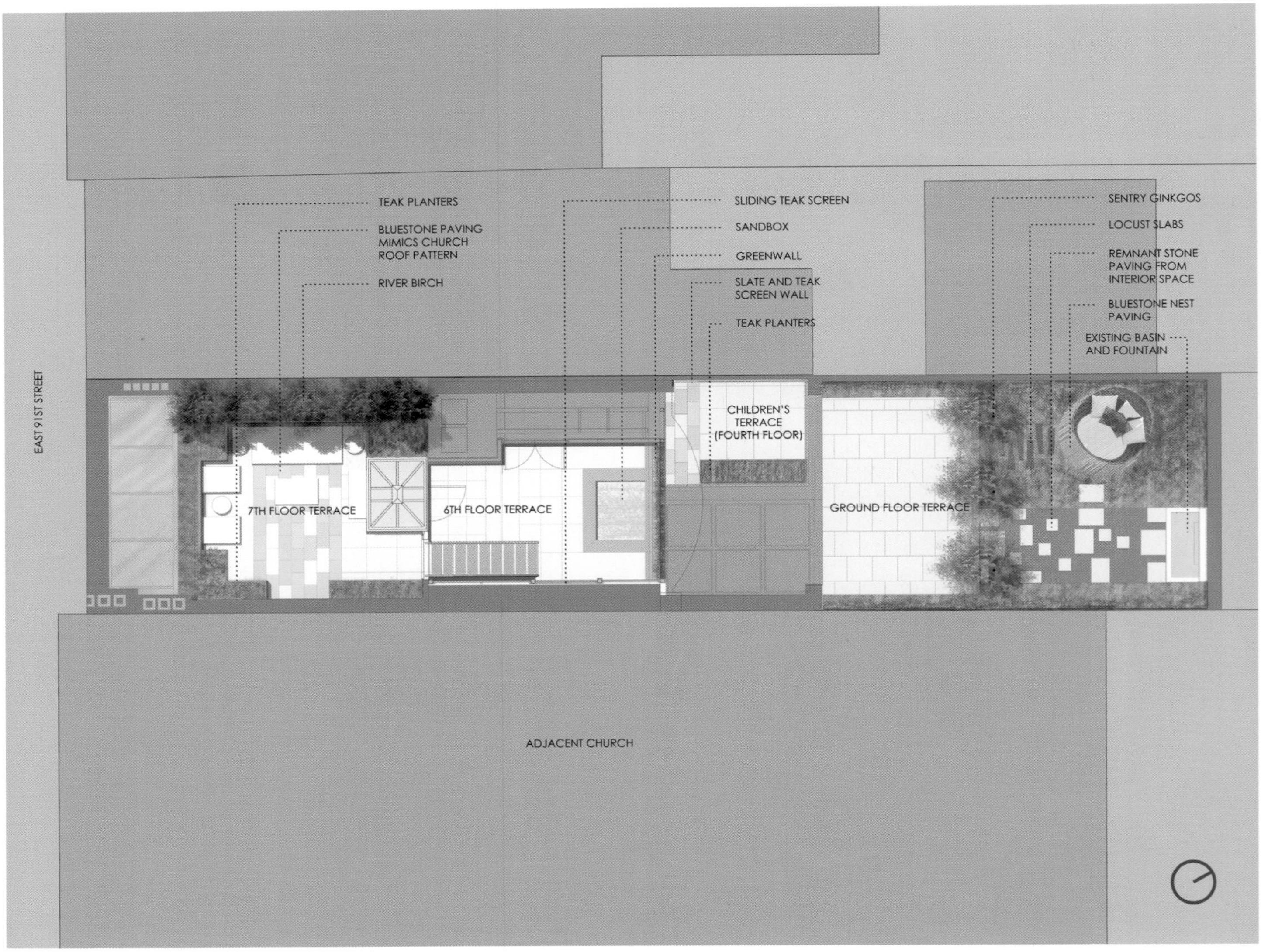